PRIME NUMBERS

The Most Mysterious Figures in Math

David Wells

WILEY

John Wiley & Sons, Inc.

Published by John Wiley & Sons, Inc., Hoboken, New Jersey
Published simultaneously in Canada

For general information about our other products and services, please contact our
Customer Care Department within the United States at (800) 762-2974, outside the
United States at (317) 572-3993 or fax (317) 572-4002.

Wiley also publishes its books in a variety of electronic formats. Some content that
appears in print may not be available in electronic books. For more information
about Wiley products, visit our web site at www.wiley.com.

Library of Congress Cataloging-in-Publication Data:

Wells, D. G. (David G.)
    Prime numbers: the most mysterious figures in math / David Wells.
        p. cm.
    Includes bibliographical references and index.
    ISBN-13 978-0-471-46234-7 (cloth)
    ISBN-10 0-471-46234-9 (cloth)
        1. Numbers, Prime. I. Title.

QA246.W35 2005
512.7'23—dc22   2004019974

Printed in the United States of America

10 9 8 7 6 5 4 3 2 1

Contents

Acknowledgments

I am delighted to thank, once again, David Singmaster for his assistance and the use of his library: on this occasion I can also note that his thesis supervisor was D. H. **Lehmer**. I am happy to acknowledge the following permissions:

The American Mathematical Society for permission to reproduce, slightly modified, the illustration on page 133 of prime knots with seven crossings or less from Pasolov and Sossinsky (1997), *Knots, links, braids and 3-manifolds*, Translations of Mathematical Monographs 154:33, Figure 3.13.

Chris Caldwell for permission to reproduce, slightly modified, the graph on page 156 showing the Mersenne primes, from his Prime Pages Web site.

The graph on page 184 comparing various historical estimates of the values of $\pi(n)$ is in the public domain, but I am happy to note that it is adapted from the diagram on page 224 of Beiler (1966), *Recreations in the Theory of Numbers*, published by Dover Publications.

Author's Note

Terms in **bold**, throughout the book, refer to entries in alphabetical order, or to entries in the list of contents, and in the index.

Throughout this book, the word *number* will refer to a positive integer or whole number, unless stated otherwise.

Letters stand for integers unless otherwise indicated.

Notice the difference between the decimal point that is on the line, as in $\frac{1}{8} = 0.125$, and the dot indicating multiplication, above the line:

$$20 = 2 \cdot 2 \cdot 5$$

Divisor and factor: these are almost synonymous. Any differences are purely conventional. As Hugh Williams puts it, if a divides b, then "we call a a *divisor* (or *factor*) of b. Since 1 and a are always divisors of a, we call these factors the *trivial* divisors (or factors) of a." (Williams 1998, 2)

On the other hand, we always talk about the prime factorization of a number, because no word like *divisorization* exists! For this reason, we also talk about finding the factors of a large number such as $2^{31} - 1$.

Similarly, by convention, the divisor function $d(n)$, which is the number of divisors of n, is never called the factor function. And so on.

The meanings of $\phi(n)$, $\sigma(n)$, and $d(n)$ are explained in the glossary.

The natural logarithm of n, the log to base e, is written as $\log n$. This does *not* mean the usual logarithm to base 10, which would be written $\log_{10} n$.

The expression $8 > 5$ means that 8 is greater than 5. Similarly, $5 < 8$ means that 5 is less than 8.

The expression $n \geq 5$ ($5 \leq n$) means that n is greater than or equal to 5 (5 is less than or equal to n).

The expression $4 \mid 12$ means that 4 divides 12 exactly.
The expression $4 \nmid 13$ means that 4 does *not* divide 13 exactly.

Finally, instead of saying, "When 30 is divided by 7 it leaves a remainder 2," it is much shorter and more convenient to write,

$$30 \equiv 2 \;(\text{mod } 7)$$

This is a congruence, and we say that "30 is congruent to 2, mod 7." The expression *mod* stands for *modulus*, because this is an example of modular arithmetic. The idea was invented by that great mathematician **Gauss**, and is more or less identical to the clock arithmetic that many readers will have met in school.

In clock (or modular) arithmetic you count and add numbers as if going around a clockface. If the clockface goes from 1 to 7 only, then 8 is the same as 1, 9 = 2, 10 = 3, and so on.

If, however, the clockface goes from 1 to 16 (for example), then 1 = 17, 2 = 18, and $3 \cdot 9 = 11$.

If you count in (say) 8s around the traditional clockface showing 12 hours, then your count will go: 8, 4, 12, 8, 4, 12, repeating endlessly and missing all the hours *except* 4, 8, and 12. If you count in 5s, however, it goes like this: 5, 10, 3, 8, 1, 6, 11, 4, 9, 2, 7, 12, 5, and by the time you start to repeat you have visited every hour on the clock. This is because 8 and 12 have a common factor 4, and but 5 and 12 have no common factor.

Mathematicians use the $\equiv$ sign instead of =, the equal sign, to indicate that they are using modular arithmetic. So instead of saying that prime numbers are always of the form $6n + 1$ or $6n - 1$, because $6n + 2$ and $6n + 4$ are even and $6n + 3$ is divisible by 3, we can write $6n \equiv \pm 1 \;(\text{mod } 6)$.

Most statements made in this book have no reference. Either they are well-known, or they can be found in several places in the literature. Even if I do know where the claim was first made, a reference is not necessarily given, because this is a popular book, not a work of scholarship.

However, where a result appears to be due to a specific author or collaboration of authors and is not widely known, I have given their names, such as (Fung and Williams). If a date is added, as in (Fung and Williams 1990), that means the reference is in the bibliography. If this reference is found in a particular book, it is given as (Fung and Williams: Guy).

The sequences with references to "Sloane" and an A number are taken from Neil Sloane's *On-Line Encyclopedia of Integer Sequences*, at www.research.att.com/~njas/sequences. See also the entry in this book for Sloane's *On-Line Encyclopedia of Integer Sequences*, as well as the "Some Prime Web Sites" section at the end of the bibliography.

The index is very full, but if you come across an expression such as $\phi(n)$ and want to know what it means, the glossary starting on page 251 will help.

Introduction

Prime numbers have always fascinated mathematicians. They appear among the integers seemingly at random, and yet not quite: there seems to be some order or pattern, just a little below the surface, just a little out of reach.

—Underwood Dudley (1978)

Small children when they first go to school learn that there are two things you can do to numbers: add them and multiply them. Addition sums are relatively easy, and addition has nice simple properties: 10 can be written as the sum of two numbers to make this pretty pattern:

$$10 = 1 + 9 = 2 + 8 = 3 + 7 = 4 + 6 =$$
$$5 + 5 = 6 + 4 = 7 + 3 = 8 + 2 = 9 + 1$$

It is also easy to write even large numbers, like 34470251, as a sum: 34470251 = 34470194 + 57. The inverse of addition, subtraction, is pretty simple also.

Multiplication is much trickier, and its inverse, division, is really quite hard; the simple pattern disappears, and writing 34470251 as a *product* is, well, fiendishly difficult. Suddenly, simple arithmetic has turned into difficult mathematics!

The difficulty is easy to understand but hard to resolve. The fact is that some numbers, the *composite* numbers, can be written as a product of two other numbers, as we learn from our multiplication tables. These numbers start with: 2×2 is 4, 2×3 is 6, and 2×4 is 8, followed later by 3×3 is 9 and 6×7 is 42, and so on.

Other numbers cannot be written as a product, except of themselves and 1. For example, $5 = 5 \times 1 = 1 \times 5$, but that's all. These are the mysterious *prime* numbers, whose sequence starts,

$$2, 3, 5, 7, 11, 13, 17, 19, 23, 29, 31,$$
$$37, 41, 43, 47, 53, 59, 61, 67, 71, \ldots$$

Notice that 1 is an exception: it is not counted as a prime number, nor is it composite. This is because many properties of prime numbers are easier to state and have fewer exceptions if 1 is *not* prime. (Zero also is neither prime nor composite.)

The prime numbers seem so irregular as to be random, although they are in fact determinate. This mixture of almost-randomness and pattern has enticed mathematicians for centuries, professional and amateur alike, to make calculations, spot patterns, make conjectures, and then (attempt) to prove them.

Sometimes, their conjectures have been false. So many conjectures about primes are as elegant as they are simple, and the temptation to believe them, to believe that you have discovered a *pattern* in the primes, can be overwhelming—until you discover the counterexample that destroys your idea. As Henri Poincaré wrote, "When a sudden illumination invades the mathematicians's mind, . . . it sometimes happens . . . that it will not stand the test of verification . . . it is to be observed almost always that this false idea, if it had been correct, would have flattered our natural instincts for mathematical elegance." (Poincaré n.d.)

Sometimes a conjecture has only been proved many years later. The most famous problem in mathematics today, by common consent, is a conjecture, the **Riemann hypothesis**, which dates from a brilliant paper published in 1859. Whoever finally proves it will become more famous than Andrew Wiles, who was splashed across the front pages when he finally proved **Fermat's Last Theorem** in 1994.

This fertility of speculation has given a special role to the modern electronic computer. In the good old bad old days, "computer" actually meant a person who computed, and a long and difficult task it could be for the mathematician who was not a human calculator like **Euler** or **Gauss**.

Today, computers can generate data faster than it can be read, and can complete calculations in seconds or hours that would have taken a human calculator years—and the computer makes no careless mistakes. (The programmer may err, of course!) Computers also put you in touch with actual numbers, in a way that an abstract proof does not. As John Milnor puts it:

> If I can give an abstract proof of something, I'm reasonably happy. But if I can get a concrete, computational proof and actually produce numbers I'm much happier. I'm rather an addict at doing things on the computer. . . . I

have a visual way of thinking, and I'm happy if I can see a picture of what I'm working with. (Bailey and Borwein 2000)

It has even been seriously argued that mathematics is becoming more of an experimental science as a result of the computer, in which the role of proof is devalued. That is nonsense: it is only by penetrating below the surface glitter that mathematicians gain the deepest understanding. Why did Gauss publish six proofs of the law of **quadratic reciprocity** (and leave a seventh among his papers)? Because each proof illuminated the phenomenon from a different angle and deepened his understanding.

Computers have had two other effects. The personal computer has encouraged thousands of amateurs to get stuck in and to explore the prime numbers. The result is a mass of material varying from the amusing but trivial to the novel, serious, and important.

The second effect is that very complex calculations needed to prove that a large number is prime, or to find its factors, have suddenly become within reach. In 1876 Édouard **Lucas** proved that $2^{127} - 1$ is prime. It remained the largest known prime of that form until 1951. Today, a prime of this size can be proved prime in a few seconds, though the problem of factorization remains intractable for large numbers, so **public key encryption** and methods such as the **RSA algorithm** have recently made prime numbers vitally important to business (and the military).

Despite the thousands of mathematicians working on properties of the prime numbers, numerous conjectures remain unresolved. Computers are wonderful at creating data, and not bad at finding counterexamples, but they prove nothing. Many problems and conjectures about prime numbers will only be eventually solved through deeper and deeper insight, and for the time being seem to be beyond our understanding. As Gauss put it, "It is characteristic of higher arithmetic that many of its most beautiful theorems can be discovered by induction with the greatest of ease but have proofs that lie anywhere but near at hand and are often found only after many fruitless investigations with the aid of deep analysis and lucky combinations." See our entry on **zeta mysteries: the quantum connection**! Gauss added, referring to his own methods of working as well as those of Fermat and Euler and others:

[I]t often happens that many theorems, whose proof for years was sought in vain, are later proved in many different ways. As soon as a new result is discovered by induction, one must consider as the first requirement the finding of a proof by *any possible* means [emphasis added]. But after such good fortune, one must not in higher arithmetic consider the investigation closed or view the search for other proofs as a superfluous luxury. For sometimes one does not at first come upon the most beautiful and simplest proof, and then it is just the insight into the wonderful concatenation of truth in higher arithmetic that is the chief attraction for study and often leads to the discovery of new truths. For these reasons the finding of new proofs for known truths is often at least as important as the discovery itself. (Gauss 1817)

The study of the primes brings in every style and every level of mathematical thinking, from the simplest pattern spotting (often misleading, as we have noted) to the use of statistics and advanced counting techniques, to scientific investigation and experiment, all the way to the most abstract concepts and most subtle proofs that depend on the unparalleled insight and intuitive perceptions of the greatest mathematicians. Prime numbers offer a wonderful field for exploration by amateurs and professionals alike.

This is not a treatise or an historical account, though it contains many facts, historical and otherwise. Rather, it is an introduction to the fascination and beauty of the prime numbers. Here is an example that I have occasionally used to, successfully, persuade nonbelievers with no mathematical background that mathematics can indeed be delightful. First write down the square numbers, $1 \cdot 1 = 1$, $2 \cdot 2 = 4$, $3 \cdot 3 = 9$, and so on. (Notice that to avoid using the $\times$ for multiplication, because x is also used in algebra, we use a dot above the text baseline.)

$$1 \quad 4 \quad 9 \quad 16 \quad 25 \quad 36 \quad 49 \quad 64 \quad 81 \quad 100 \quad \ldots$$

This sequence is especially simple and regular. Indeed, we don't even need to multiply any numbers to get it. We could just as well have started with 1 and added the odd numbers. $1 + 3 = 4$; $4 + 5 = 9$; $9 + 7 = 16$, and so on.

Now write down the prime numbers, the numbers with no factors except themselves and 1:

$$2 \quad 3 \quad 5 \quad 7 \quad 11 \quad 13 \quad 17 \quad 19 \quad 23 \quad 29 \quad \ldots$$

No such simplicity here! The jumps from one number to the next vary irregularly from 1 to 6 (and would eventually become much larger). Yet there is a concealed pattern connecting these two sequences. To see it, strike out 2, which is the only even prime, and all the primes that are *one less* than a multiple of 4; so we delete 3, 7, 11, 19, and 23 . . . The sequence of remaining primes goes,

$$5 \quad 13 \quad 17 \quad 29 \quad 37 \quad 41 \quad 53 \quad 57 \quad 61 \quad 73 \quad \ldots$$

And the connection? Every one of these primes is the *sum of two squares*, of two of the numbers in the first sequence, in a unique way:

$$5 = 1 + 4, \quad 13 = 4 + 9, \quad 17 = 1 + 16, \quad 29 = 4 + 25, \quad 37 = 1 + 36$$

and so on. This extraordinary fact is related to Pythagoras's theorem about the sides of a right-angled triangle, and was known to **Diophantus** in the third century. It was explored further by **Fermat**, and then by **Euler** and **Gauss** and a host of other great mathematicians. We might justly say that it has been the mental springboard and the mysterious origin of a large portion of the theory of numbers—and yet the basic facts of the case can be explained to a school pupil.

There lies the fascination of the prime numbers. They combine the maximum of simplicity with the maximum of depth and mystery. On a plaque attached to the NASA deep space probe we are described in symbols for the benefit of any aliens who might meet the spacecraft as "bilaterally symmetrical, sexually differentiated bipeds located on one of the outer spirals of the Milky Way, capable of recognizing the prime numbers and moved by one extraordinary quality that lasts longer than all our other urges—curiosity."

I hope that you will discover (or be reminded of) some of the fascination of the primes in this book. If you are hooked, no doubt you will want to look at other books—there is a selection of recommended books marked in the bibliography with an asterisk—and you will also find a vast amount of material on the Internet: some of the best sites are listed at the "Some Prime Web Sites" section at the end of the bibliography. To help you with your own research, Appendix A is a list of the first 500 primes, and Appendix B lists the first 80 values of the most common arithmetic functions.

Note: As this book went to press, the record for the largest known prime number was broken by Dr. Martin Nowak, a German eye specialist who is a member of the worldwide **GIMPS** (Great Internet Mersenne Prime Search) project, after fifty days of searching on his 2.4GHz Pentium 4 personal computer. His record prime is $2^{25,964,951} - 1$ and has 7,816,230 digits.

abc conjecture

The abc conjecture was first proposed by Joseph Oesterlé and David Masser in 1985. It concerns the product of all the *distinct* prime factors of n, sometimes called the *radical* of n and written $r(n)$. If n is **square-free** (not divisible by any perfect square), then $r(n) = n$. On the other hand, for a number such as $60 = 2^2 \cdot 3 \cdot 5$, $r(60) = 2 \cdot 3 \cdot 5 = 30$.

$r(n)$ is smallest when n is a power of a prime: then $r(p^q) = p$. So $r(8) = r(32) = r(256) = 2$, and $r(6561) = r(3^8) = 3$.

The more duplicated factors n has, the larger n will be compared to $r(n)$. For example, if $n = 9972 = 2^2 \cdot 3^2 \cdot 277$, then $r(9972) = 1662$, and $r(n) = \frac{1}{6}n$.

The abc conjecture says, roughly, that if a and b are two numbers with no common factor, and sum c, then the number abc cannot be very composite. More precisely, David Masser proved that the ratio $r(abc)/c$ can be as small as you like. Less than $\frac{1}{100}$? Yes! Less than 0.00000001? Yes! And so on.

However—and this is Masser's claim and the abc conjecture—this is *only just possible*. If we calculate $r(abc)^n/c$ instead, where n is *any* number greater than 1, then we can't make $r(abc)/c$ as small as we like, and this is true even if n is only slightly greater than 1. So even if n is as small as 1.00001, $r(abc)^n/c$ has a lower limit that isn't zero.

Why is this conjecture about numbers that are not squarefree so important? Because, incredibly, so many important theorems could be proved quite easily, *if it were true*. Here are just five of the many consequences of the abc conjecture being true:

- **Fermat's Last Theorem** could be proved very easily. The proof by Andrew Wiles is extremely long and complex.
- There are infinitely many **Wieferich primes**.
- There is only a finite number of sets of three consecutive **powerful numbers**.
- There is only a finite number of solutions satisfying **Brocard's equation**, $n! + 1 = m^2$.
- All the polynomials $(x^n - 1)/(x - 1)$ have an infinity of square-free values. (Browkin 2000, 10)

abundant number

A number is abundant if the sum of its *proper* divisors (or aliquot parts, meaning all its divisors except the number itself) is greater than the number. Roughly speaking, numbers are abundant when they have several different small prime factors. Thus $12 = 2^2 \cdot 3$ is abundant, because $1 + 2 + 3 + 4 + 6 = 16 > 12$.

Abundant numbers were presented by Nicomachus (c. AD 100) in his *Introduction to Arithmetic*, which included definitions of prime numbers (he did not consider 1, or unity, and 2 to be numbers) and also **deficient** and **perfect** numbers, explaining that,

> Among simple even numbers, some are superabundant, others are deficient: these two classes are as two extremes opposed to one another; as for those that occupy the middle position between the two, they are said to be perfect. And those which are said to be opposite to each other, the superabundant and the deficient, are divided in their condition, which is inequality, into the too much and the too little.
>
> In the case of the too much, is produced excess, superfluity, exaggerations and abuse; in the case of too little, is produced wanting, defaults, privations and insufficiencies. And in the case of those that are found between the too much and the too little, that is in equality, is produced virtue, just measure, propriety, beauty and things of that sort—of which the most exemplary form is that type of number which is called perfect. (O'Connor and Robertson n.d.)

He also wrote, in the style of the period, that "even abundant numbers" are like an animal with "too many parts or limbs, with ten tongues, as the poet says, and ten mouths, or with nine lips, or three rows of teeth," whereas perfect numbers are linked to "wealth, moderation, propriety, beauty, and the like." (Lauritzen, Versatile Numbers)

Nicomachus claimed that all odd numbers are deficient. Most abundant numbers are indeed even. The smallest odd abundant is $945 = 3^3 \cdot 5 \cdot 7$. There are only twenty-three odd abundant numbers less than 10,000.

Every multiple of an abundant number is abundant. Therefore, there is an infinite number of abundant numbers. The sequence starts:

12   18   20   24   30   36   40   42   48   54   56   . . .

The pair 54 and 56 is the first abundant numbers with the same sum of proper divisors, 120. The next pairs are 60 and 78 (sum = 168) and 66 and 70 (sum = 144).

Roughly 24.8% of the positive integers are abundant.

The sum of all the **divisors** of n, including n itself, is called $\sigma(n)$.

When $n = 12$, $\sigma(n)/n = 28/12 = 7/3$, which is a record for numbers up to 12. Any number that sets a record for $\sigma(n)/n$ is called *super-abundant*. These are the first few record-breaking values of $\sigma(n)/n$:

n	1	2	6	12	24	36	48	60
$\sigma(n)/n$	1	3/2	2	7/3	5/2	91/36	31/12	14/5

If n is even and $\sigma(n)/n > 9$, then it has at least fifty-five distinct prime factors.

Every number greater than 20161 is the sum of two abundant numbers.
See deficient number; divisors; perfect number

AKS algorithm for primality testing

Our world resonates with patterns. The waxing and waning of the moon. The changing of the seasons. The microscopic cell structure of all living things have patterns. Perhaps that explains our fascination with prime numbers which are uniquely without pattern. Prime numbers are among the most mysterious phenomena in mathematics.

—*Manindra Agrawal (2003)*

The ideal **primality test** is a definite yes-no test that also runs quickly on modern computers. In August 2002, Manindra Agrawal of the Indian Institute of Technology in Kanpur, India, and his two brilliant PhD students Neeraj Kayal and Nitin Saxena, who were both in the 1997 Indian Mathematics Olympiad Squad, announced just such a test, using his own novel version of **Fermat's Little Theorem**, in a short paper of only nine pages that was also extremely simple and elegant.

In a sign of the times, Agrawal sent an e-mail to a number of prominent mathematicians with the subject header "PRIMES is in P," and also put it on his Web site. It was downloaded more than thirty thousand times in the first twenty-four hours, and the site was visited more than two million times in the first ten days. (Earlier, AKS had reached a gap in their attempted proof, which they filled by searching the Web and finding just the mathematical result they needed.)

"PRIMES is in P" means that a number can be tested to decide whether or not it is prime in a time that is roughly proportional to its number of digits. This means that it is fast for very large numbers but not so fast for the kind of numbers that often have to be tested in practical applications. Fortunately, in another sign of the times, within hours of its publication other mathematicians were finding variations on the original AKS algorithm that made it much faster. Currently, the most-improved versions will run about two million times faster. This nearly makes it competitive with the most efficient current algorithm— but Agrawal will never benefit financially, because he decided against trying to patent the result.

The algorithm is so simple that it has prompted many mathematicians to wonder what other problems might have unexpectedly simple solutions: for example, the problem of factorizing large numbers. Agrawal's algorithm is no help here: the most it can do is show that a number is composite, without saying anything about its factors, so it will have no effect on **encryption** using prime numbers. (Agrawal 2002)

See primality testing

aliquot sequences (sociable chains)

The *aliquot parts* (the expression is old-fashioned) of a number are its proper **divisors**, meaning its divisors apart from the number itself.

Any integer is the start of an aliquot sequence. Simply calculate the sum of its proper divisors and then repeat. Starting with 10 we soon reach 1: the proper divisors of 10 are 1, 2, and 5, summing to 8; of 8 they are 1, 2, and 4, summing to 7, which is prime, so its only proper divisor is 1.

For 24 we get this sequence:

$$24 \quad 36 \quad 55 \quad 17 \quad 1$$

However, 28 immediately repeats, because $1 + 2 + 4 + 7 + 14 = 28$, and so 28 is a **perfect number**, while 220 and 284 each lead at once to the other, so they form an **amicable** pair.

For reasons that are not understood, many aliquot sequences end in Paganini's amicable pair, 1184 and 1210.

The third possibility is that the sequence repeats through a cycle; the first two examples of such sociable chains or aliquot cycles were found by Poulet in 1918. The smaller is:

12496   14288   15472   14536   14264

The second chain is of twenty-eight numbers: 14316, 19116, 31704, 47616, 83328, 177792, 295488, 629072, 589786, 294896, 358336, 418904, 366556, 274924, 275444, 243760, 376736, 381028, 285778, 152990, 122410, 97946, 48976, 45946, 22976, 22744, 19916, 17716, (14316). It is remarkable how little oscillation there is in this sequence. Drawn as a graph there would be just four peaks, at 629072, 418904, 275444, and 381028. (Beiler 1966, 29)

No more chains were discovered until 1969 when Henri Cohen checked all aliquot sequences starting under 60,000,000 and found seven chains of four links each. No chain of three links—nicknamed a "crowd"!—has ever been found, though no one has a reason why they should not exist.

Catalan in 1888 and then Dickson conjectured that no aliquot sequence goes off to infinity—they all end in a cycle or in 1. A sequence starting with an abundant number will initially increase; however, there are far more deficient than abundant numbers, which suggests that most sequences will indeed decrease more than increase.

There are just seventeen numbers less than 2000 for which the problem is unsolved: 276, 552, 564, 660, 966, 1074, 1134, 1464, 1476, 1488, 1512, 1560, 1578, 1632, 1734, 1920, and 1992. Notice that they are all even. It has been conjectured that the aliquot sequences for most even numbers do not end in 1 or a cycle.

The first five of these numbers are the so-called Lehmer five. Originally the list was the Lehmer six, but then the fate of 840 was settled. It eventually reaches 1, after peaking at:

3 463982 260143 725017 429794 136098 072146 586526 240388

D. N. Lehmer showed that 138 rises after 117 steps to 179,931,895,322 and then ends in 1 after 177 steps. (Guy 1981, B6)

Wolfgang Creyaufmueller found the longest known terminating sequence in 2002. It starts at 446580 and ends 4,736 steps later with the prime 601, followed by 1. (Creyaufmueller 2002)

Manuel Benito and Juan Varona found the sequence with the highest known peak: it starts with 3630 and has a maximum length of 100 digits, ending after 2,624 steps with the prime 59, and then 1. (Benito and Varona 2001)

almost-primes

The almost-prime numbers have a limited number of prime factors. The 2-almost-primes have two prime factors (including duplicated factors) and are also called **semiprimes**: the 3-almost-primes have three, and so on.

The sequence of 3-almost-primes starts 8, 12, 18, 20, 27, 28, 30, 42, 44, 45, 50, . . .

The sequence of n-almost-primes starts with 2^n, $3 \cdot 2^{n-1}$, . . .

amicable numbers

A pair of numbers is amicable (or semiperfect) if each is the sum of the proper divisors of the other. The smallest pair is 220 and 284. The proper divisors of $220 = 2^2 \cdot 5 \cdot 11$ sum to,

$$1 + 2 + 4 + 5 + 10 + 11 + 20 + 22 + 44 + 55 + 110 = 284$$

and similarly: $284 = 2^2 \cdot 71$ and $1 + 2 + 4 + 71 + 142 = 220$.

According to the philosopher Iamblichus (c. AD 250–330), the followers of Pythagoras "call certain numbers amicable numbers, adopting virtues and social qualities to numbers, such as 284 and 220; for the parts of each have the power to generate the other," and Pythagoras described a friend as "one who is the other I, such as are 220 and 284."

In the Bible (Genesis 32:14), Jacob gives 220 goats (200 female and 20 male) to Esau on their reunion. There are other biblical references at Ezra 8:20 and 1 Chronicles 15:6, while 284 occurs in Nehemiah 11:18. These references are all to the tribe of Levi, whose name derives from the wish of Levi's mother to be amicably related to his father. (Aviezri and Fraenkel: Guy 1994)

They were also used in magic and astrology. Ibn Khaldun (1332–1406) wrote that "the art of talismans has also made us recognize the marvelous virtues of amicable (or sympathetic) numbers. These numbers are 220 and 284. . . . Persons who occupy themselves with talismans assure that these numbers have a particular influence in establishing union and friendship between individuals." (Ore 1948, 97)

Thabit ibn Qurra (c. AD 850) in his *Book on the Determination of Amicable Numbers* noted that if you choose n so that each of the expressions $a = 3 \cdot 2^n - 1$, $b = 3 \cdot 2^{n-1} - 1$, and $c = 9 \cdot 2^{2n-1} - 1$ is prime, then $2^n ab$ and $2^n c$ are amicable numbers. Unfortunately, it isn't easy to make them all prime at once, and in fact it only works for $n = 2$, 4, and 7 and no other n less than 20,000.

A second pair, 17,296 and 18,416, was discovered by Ibn al-Banna (1256–1321) and rediscovered by **Fermat** in 1636. Descartes found the third pair, 9,363,584 and 9,437,056, which is the case $n = 7$ in Thabit's formulae. **Euler** then discovered no less than sixty-two more examples, without following Thabit's rule.

Paganini's amicable pair, 1184 and 1210, is named after Nicolo Paganini, who discovered them in 1866 when he was a sixteen-year-old schoolboy. They had previously been missed by Fermat, Descartes, Euler, and others.

More than 7,500 amicable pairs have been found, using computers, including all pairs up to 10^{14}. Is there an infinite number of amicable pairs? It is generally believed so, partly because Herman te Riele has a method of constructing "daughter" pairs from some "mother" pairs. Te Riele has also published all of the 1,427 amicable pairs less than 10^{10}.

X	no. of pairs with smaller no. $< X$
10^3	1
10^4	5
10^5	13
10^6	42
10^7	108
10^8	236
10^9	586
10^{10}	1427

(Gupta)

amicable curiosities

- There is no known amicable pair in which one number is a square.
- The numbers in amicable pairs end in 0 or 5 surprisingly often, for no known reason.

- Most amicable numbers have many different factors. Can a power of a prime, p^n, be one of an amicable pair? If so, then $p^n > 10^{1500}$ and $n > 1400$.
- It is not known whether there is a pair of coprime amicable numbers. If there is, the numbers must exceed 10^{25} and their product must have at least twenty-two distinct prime factors.

Andrica's conjecture

Dorin Andrica conjectured that $\sqrt{p_{n+1}} - \sqrt{p_n} < 1$ for all n. This is really a conjecture about the gaps between prime numbers and is not even a very strong conjecture, yet it has never been proved. The largest value of the difference for n less than 1000 is $\sqrt{11} - \sqrt{7} = 0.670873\ldots$ which is well below 1.

Imran Ghory has used data on largest prime gaps to confirm the conjecture up to $1.3002 \cdot 10^{16}$.

arithmetic progressions, of primes

In an arithmetic progression (or sequence) the differences between successive terms are constant, for example:

$$3 \quad 7 \quad 11 \quad 15 \quad 19 \quad 23 \quad 27 \quad 31 \quad 35 \quad 39 \quad 43 \quad \cdots$$

with constant difference 4. This happens to already contain seven primes, with one sequence of three consecutive primes.

The current record for the largest number of *consecutive* primes in arithmetic progression has ten primes. It was set 11:56 a.m. on March 2, 1998, by Manfred Toplic of Klagenfurt, Austria, in a typical example of **distributed computing**. The first term is the prime 100 99697 24697 14247 63778 66555 87969 84032 95093 24689 19004 18036 03417 75890 43417 03348 88215 90672 29719, and the common difference is 210.

The same team also set the previous record of nine consecutive primes, on January 15, 1998. The team was led by Harvey Dubner and Tony Forbes. More than seventy people, using about two hundred machines, searched nearly fifty ranges of a trillion numbers each.

The longest known arithmetic progression of nonconsecutive primes was discovered by Pritchard, Moran, and Thyssen in 1993. It is twenty-two terms long, starting with the prime 11410337850553 and with common difference 4609098694200. On April 22, 2003, another twenty-two-term sequence was found by Markus Frind.

The largest triple of primes in arithmetic progression is the 13,447-digit sequence starting $475977645 \cdot 2^{44640} - 1$ with common difference $475977645 \cdot 2^{44639} - 2$, discovered by Herranen and Gallot in 1998.

The largest quadruple of primes in arithmetic progression is the 1,815-digit sequence starting $174499605 \cdot 2^{6000} + 1$ with common difference $20510280 \cdot 2^{6000}$, found by Roonguthai and Gallot in 1999.

The set of smallest prime progressions starts:

no. of terms	minimum difference	smallest progression
2	1	2, 3
3	2	3, 5, 7
4	6	5, 11, 17, 23
5	6	5, 11, 17, 23, 29
6	30	7, 37, 67, 97, 127, 157
7	150	7, 157, 307, . . .
8	210	199, 409, 619, . . .
9	210	199, 409, 619, . . .
10	210	199, 409, 619, . . .
11	2310	60858179, . . .

The longest known arithmetic progression of primes is twenty-two terms long, starting from 11,410,337,850,553 with difference 4,609,098,694,200.

In 1939 van der Corput proved that an infinity of triples of primes in arithmetic progression exists. Ben Green of the University of British Columbia and Terence Tao of the University of California at Los Angeles proved in 2004 that prime arithmetic progressions of any length do exist, though their proof, like so many proofs, is nonconstructive, so they cannot actually generate any examples.

See Dickson's conjecture; Dirichlet; Hardy-Littlewood conjectures

Aurifeuillian factorization

Since $a^2 + b^2$ cannot be factorized into two algebraic factors, unlike $a^2 - b^2 = (a + b)(a - b)$, we might assume that $n^4 + 1$, which is also the sum of two squares, cannot be factorized. Not so!

$$n^4 + 1 = (n^2 - n + 1)(n^2 + n + 1)$$

Now we can see a connection: $a^2 + b^2 = (a + b)^2 - 4ab = (a - \sqrt{ab} + b)(a + \sqrt{ab} + b)$, which normally "doesn't count" because of the square roots. It follows that $n^4 + 1$ is always composite, except when $n^2 - n + 1 = 1$ and $n = 0$ or 1.

This is an example of an *Aurifeuillian* factorization, named after Léon François Antoine Aurifeuille, who discovered a special case in 1873:

$$2^{4m-2} + 1 = (2^{2m-1} + 2^m + 1)(2^{2m-1} - 2^m + 1)$$

Knowledge of this factorization would have saved the many years of his life that Fortuné Landry spent factoring $2^{58} + 1$, finally finishing in 1869. Landry's gargantuan factorization is just a trivial special case!

$$2^{58} + 1 = (2^{29} + 2^{15} + 1)(2^{29} - 2^{15} + 1)$$

Édouard **Lucas** later found more Aurifeuillian factorizations, which are related to the complex roots of unity. Here are two more examples:

$$3^{6k-3} + 1 = (3^{2k-1} + 1)(3^{2k-1} - 3^k + 1)(3^{2k-1} + 3^k + 1)$$

$$5^{5b} - 1 = (5^b - 1)LM, \text{ where } L = T^2 - T5^k + 5^b \text{ and}$$
$$M = T^2 + T5^k + 5^b \text{ and } T = 5^b + 1, b = 2k - 1.$$

Aurifeuillian factors have other uses. For example, if L_n is the nth **Lucas number**, and n is odd, then

$$L_{5n} = L_n A_{5n} B_{5n} \text{ where } A_{5n} = 5F_n^2 - 5F_n + 1 \text{ and } B_{5n} = 5F_n^2 + 5F_n + 1$$

average prime

If $S(k)$ is the sum of the first k prime numbers, then the average of the first k primes is $S(k)/k$. This is an integer for these values of k:

k	p_k	$S(p_k)$	$S(p_k)/k$
1	2	2	2
23	83	874	38
53	241	5830	110
853	6599	2615298	3066
11869	126551	712377380	60020
117267	154479	86810649294	740282
339615	4864121	794712005370	2340038
3600489	60686737	105784534314378	29380602
96643287	1966194317	92542301212047102	957565746

(Rivera, Puzzle 31)

Bang's theorem

Does every term in a sequence contain at least one prime factor that has not appeared before in the sequence? Such a prime factor is called **primitive**.

If $a > 1$ is fixed, then every number $a^n - 1$ has a primitive prime factor, with the sole exception of $2^6 - 1 = 63$. Similarly, if $a > 1$, then every number $a^n + 1$ has a primitive prime factor, with the sole exception of $2^3 + 1 = 9$. This was proved by Bang in 1886, and incidentally offers another way to prove that there is an infinity of prime numbers.

Zsigmondy proved the same theorem for the more general functions $a^n - b^n$ and $a^n + b^n$, with the same condition and the same exceptions. The sequence for $T = 2^n + 3^n$ starts:

n	1	2	3	4	5	6	7	8	9	10
T	5	13	35	97	275	793	2315	6817	20195	60073
	5	13	$5 \cdot 7$	97	$5^2 \cdot 11$	$13 \cdot 61$	$5 \cdot 463$	$17 \cdot 401$	$5 \cdot 7 \cdot 577$	$13 \cdot 4621$

Bateman's conjecture

$$1 + 2 + 2^2 + 2^3 + 2^4 = 1 + 5 + 5^2 = 31$$

Is this the only sum of this kind, using *prime* numbers? No one knows. If composite numbers are allowed, there is at least one other solution:

$$1 + 2 + 2^2 + 2^3 + \ldots + 2^{12} = 1 + 90 + 90^2 = 8191$$

Beal's conjecture, and prize

The Texas millionaire Andrew Beal, the fifty-one-year-old founder of the Beal Bank and Beal Aerospace Technologies that builds rockets for satellite launches, and a number enthusiast, is offering a reward to the first person to prove (or disprove) this conjecture, which is a generalization of **Fermat's Last Theorem**:

If $x^m + y^n = z^r$ where x, y, z, m, n, and r are all positive integers, and m, n, and r are greater than 2, then x, y, and z have a common factor.

Without the condition that m, n, and r must be greater than 2, there are many solutions, including all Pythagorean triples starting with $3^2 + 4^2 = 5^2$ and $5^2 + 12^2 = 13^2$, and the solutions to the **Fermat-Catalan conjecture**. It follows, from a theorem of Falting, that for any particular choice of m, n, and r, there can only be a finite number of solutions, but are there any at all?

The conjecture and prize were originally announced in 1997 in the prestigious *Notices of the American Mathematical Society*, originally with a prize of $5,000 rising by $5,000 a year to a maximum of $50,000. Since then the prize has been increased to $100,000 for either a proof or a counterexample. The prize money has been handed to the American Mathematical Society for safekeeping and the interest is being used to fund the annual Erdös Memorial Lecture.

Just in case anyone thinks that they can work out the answer on a scruffy piece of paper, the award will be given only when "the solution has been recognized by the mathematics community. This includes that either a proof has been given and the result has appeared in a reputable referred journal or a counterexample has been given and verified." (www.bealconjecture.com)

The solution is sure to be difficult because the conjecture is based on extensive numerical tests. Beal and a colleague spent thousands of hours searching for solutions for various values of the exponents, only to find that when solutions appeared, a pair out of x, y, and z always had a common factor. Hence the conjecture, which is surprisingly novel. (A similar but not identical idea was conjectured by Viggo Brun in 1914.)

If the **abc conjecture** is true, then there are no solutions to Beal's equation when the exponents are large enough, and Darmon and Granville showed in 1995 that in effect there are at most a finite number of solutions. But are there any?

See Fermet-Catalan equation and conjecture.

Benford's law

If numbers in general were equally likely to start with any of the digits 1 to 9, then out of the 78,498 prime numbers less than 1,000,000 we would expect about one-ninth of them to begin with the digit 1, or about 8,700, but no, there are 9,585 such primes starting with the digit 1. In fact, from first digit 1 to first digit 9, the number of primes in each category decreases.

Why this difference? Because in very many circumstances (not all) numbers begin with the digit 1 more often than with other digits. This was first noticed by the nineteenth-century astronomer Simon Newcomb, who claimed, "That the ten digits do not occur with equal frequency must be evident to anyone making use of logarithm tables, and noticing how much faster the first pages wear out than the last ones. The first significant figure is oftener 1 than any other digit and the frequency diminishes up to 9."

His conclusion was taken up again by Benford, a physicist working for the General Electric Company in 1938. He concluded that the first digit is d with probability $\log_{10}(1 + 1/d)$, which for $d = 1$ is approximately 0.30103.

initial digit	1	2	3	4	5	6	7	8	9
Benford's law	.301	.176	.125	.097	.079	.067	.058	.051	.046

These are the frequencies of first digits among the first 100 **Fibonacci numbers**, closely matching Benford's law:

initial digit	1	2	3	4	5	6	7	8	9
frequency	30	18	13	9	8	6	5	7	4

It is sometimes assumed, without any sound reason, that Benford's law is universal, that it applies to every set of numbers, anywhere, as if it were "a built-in characteristic of our number system." This isn't so. A counterexample is the powers of 2, at least for low powers. Here are the frequencies of the first digits of 2^n from $n = 0$ to 60:

digit	1	2	3	4	5	6	7	8	9
frequency	19	12	6	6	6	4	2	5	1
Benford's law	18	11	7	6	5	4	3	3	3

The match is good to start with, but then poor, with a marked spike at 8. (Raimi 1976)

Bernoulli numbers

The Bernoulli numbers are defined by this equation:

$$\frac{x}{e^x - 1} = B_0 + \frac{B_1 x}{1!} + \frac{B_2 x^2}{2!} + \frac{B_3 x^3}{3!} + \frac{B_4 x^4}{4!} + \ldots$$

The first few values are:

$B_0 = 1 \qquad B_1 = -\frac{1}{2} \qquad B_2 = \frac{1}{6} \qquad B_3 = B_5 = B_7 = \ldots = B_{2n+1} = 0$

$B_4 = -\frac{1}{30} \qquad B_6 = \frac{1}{42} \qquad B_8 = -\frac{1}{30} \qquad B_{10} = \frac{5}{66} \qquad B_{12} = -\frac{691}{2730}$

$B_{14} = \frac{7}{6} \qquad B_{16} = -\frac{3617}{510} \qquad B_{18} = \frac{43867}{798}$

Ada Lovelace and the First Computer Algorithm

In 1840 Charles Babbage asked his collaborator Ada Lovelace, daughter of Lord Byron, to add her own notes to a manuscript on his Analytical Engine. The machine used cards based on those used to control the Jacquard loom (and which were forerunners of the Holerith cards used in early modern computers).

In her notes Lovelace emphasized (as we would put it today) the interplay between programming and machinery, software and hardware:

> In enabling mechanism to combine together general symbols in successions of unlimited variety and extent, a uniting link is established between the operations of matter and the abstract mental processes of the most abstract branch of mathematical science. A new, a vast, and a powerful language is developed for the future use of analysis.

She concluded by explaining how the engine could compute the Bernoulli numbers, and made another comment that today's computer programmer will recognize at once:

> We may here remark, that the average estimate of three Variable-cards coming into use to each operation, is not to be taken as an absolutely and literally correct amount for all cases and circumstances. Many special circumstances, either in the nature of a problem, or in the arrangements of the engine under certain contingencies, influence and modify this average to a greater or less extent.

This is generally considered to be the first account of a computer algorithm. (Menabrea 1842)

Bernoulli numbers can also be calculated using the binomial coefficients from **Pascal's triangle**:

$$B_0 = 1$$
$$2B_1 + 1B_0 = 0 \qquad\qquad \text{so } B_1 = -\tfrac{1}{2}$$
$$3B_2 + 3B_1 + B_0 = 0 \qquad\qquad \text{so } B_2 = \tfrac{1}{6}$$
$$4B_3 + 6B_2 + 4B_1 + B_0 = 0 \qquad\qquad \text{so } B_3 = 0$$
$$5B_4 + 10B_3 + 10B_2 + 5B_1 + B_0 = 0 \qquad \text{so } B_4 = -\tfrac{1}{30}$$

and so on.

There is also a connection between the Bernoulli numbers and the Riemann zeta function:

$$B_n = (-1)^{n+1} n \zeta(1 - n)$$

Bernoulli number curiosities

- The denominator of B_n is always **squarefree**.
- The denominator of B_{2n} equals the product of all the primes p such that $p - 1 \mid 2n$.
- The fractional part of B_n in the decimal system has a decimal period that divides n, and there is a single digit before that period. (Conway and Guy 1996, 107–10)
- G. J. Fee and S. Plouffe have computed $B_{200,000}$, which has about 800,000 digits.

Bertrand's postulate

Joseph Bertrand (1822–1900) was a precocious student who published his first paper, on electricity, at the age of seventeen, but then became more notable as a teacher than as an original mathematician.

Bertrand's postulate states that if n is an integer greater than 3, then there is at least one prime between n and $2n - 2$. (This is the precise theorem. It is often claimed that there is a prime between n and $2n$, which is a weaker claim.)

Strangely, although it continues to be called a postulate, it is actually a theorem: it was proved by Tchebycheff in 1850 after Bertrand in 1845 had verified it for n less than 3,000,000. It is also a rather weak theorem that can be strengthened in several ways:

- Provided n is large enough, there are at least k primes between n and $2n$, however large the value of k.
- If n is at least 48, then there is at least one prime between n and $9n/8$.
- If n is greater than 6, then there is at least one prime of the form $4k + 1$ and at least one of the form $4k + 3$ between n and $2n$.
- If n is greater than or equal to 118, then the interval n to $4n/3$ inclusive contains a prime of each of the forms $4n + 1$, $4n - 1$, $6n + 1$, and $6n - 1$.
- If n is greater than 15, then there is at least one number between n and $2n$ that is the product of three different primes.

It also follows from Bertrand's postulate that:

- There is at least one prime of any given digit length beginning with the digit 1, in any base, not just base 10.
- The first $2k$ integers can always be arranged in k pairs so that the sum of the entries in each pair is a prime.
- There is a number c such that the integral parts of 2^c, 2^{2^c}, $2^{2^{2^c}}$, . . . are primes. The constant c is approximately 1.25164759777905. The first four primes are 2, 5, 37, 137438953481. The number c is not sufficiently accurately known to calculate the next prime in the sequence. (R. L. Graham, D. E. Knuth, & O. Patashnik)

Bonse's inequality

This states that if p_n is the nth prime, then

$$p_{n+1}^2 < p_1 p_2 p_3 \cdots p_n$$

provided $n > 3$.

Brier numbers

A **Riesel number** is an integer k such that $k \cdot 2^n - 1$ is composite for any integer value of n, and a **Sierpinski number** is an integer k such that $k \cdot 2^n + 1$ is composite for any integer value of n.

What about the *Brier numbers*, which are simultaneously Riesel and Sierpinski? Eric Brier was the first to find one:

29364695660123543278115025405114452910889

Yves Gallot found three smaller Brier numbers in January 2000. The smallest is twenty-seven-digits: 878503122374924101526292 469. (Rivera, Problem 29)

See Riesel number; Sierpinski numbers

Brocard's conjecture

Brocard conjectured in 1904 that the only solutions of

$$n! + 1 = m^2$$

are $n = 4$, 5, and 7. There are no other solutions with $n < 10^9$. (Berndt and Galway n.d.)

Another of Brocard's conjectures is that there are at least four primes between the squares of any two consecutive primes, with the exception of 2 and 3. This is related to Schinzel's conjecture that, provided x is greater than 8, there is a prime between x and $x + (\log x)^2$.

See Opperman's conjecture

Brun's constant

In 1919 Viggo Brun (1885–1978) proved that the sum of the reciprocals of the **twin primes** converges to Brun's constant:

$$\frac{1}{3} + \frac{1}{5} + \frac{1}{5} + \frac{1}{7} + \frac{1}{11} + \frac{1}{13} + \frac{1}{17} + \frac{1}{19} + \ldots = 1.9021605 \ldots$$

It was in 1994, while he was trying to calculate Brun's constant, that Thomas R. Nicely discovered a famous flaw in the Intel Pentium microprocessor. The Pentium chip occasionally gave wrong answers to a floating-point (decimal) division calculations due to errors in five entries in a lookup table on the chip. Intel spent millions of dollars replacing the faulty chips.

More recently, Nicely has calculated that the value of Brun's constant based on all the pairs of twin primes less than $5 \cdot 10^{15}$ is $1.902160582582 \pm 0.000000001620$. (Nicely 2004a) These are the first few approximate sums:

limit	no. of twin prime pairs	approx. sum of reciprocals
1,000	35	1.5180
10,000	205	1.6169
100,000	1,224	1.6728
1,000,000	8,169	1.7108
10,000,000	58,980	1.7384

Viggo Brun's methods have been used to study **Goldbach's conjecture** and the twin primes conjecture and to prove that there exist infinitely many integers n such that n and $n + 2$ have at most nine prime factors, and that all large even integers are the sum of two integers each having at most nine prime factors.

See Mertens constant

Buss's function

Frank Buss has defined a function, $B(n)$, that seems to generate only primes. It is calculated like this:

$$f(1) = 1$$
$$B(n) = [\text{next prime to } (f(n) + 1)] - f(n)$$
$$f(n) = f(n - 1) \cdot B(n - 1)$$

The sequence starts:

n	1	2	3	4	5	6	7
$f(n)$	1	2	6	30	210	2730	30030
"next-prime"	3	5	11	37	223	2741	30047
$B(n)$	2	3	5	7	13	11	17

The conjecture has been tested successfully up to $n = 603$. However, like so many such conjectures, it seems likely that this is a case of the the **strong law of small numbers**. (Rivera, Conjecture 29)

Carmichael numbers

According to **Fermat's Little Theorem**, if p is prime and n and p are coprime (they have no common factor), then $n^{p-1} \equiv 1 \pmod{p}$.

However, some composite numbers satisfy this equation also, and do so for *every* value of n. These are the Carmichael numbers, named after Robert Daniel Carmichael (1879–1967). They are sometimes called *absolute pseudoprimes* because they are **pseudoprimes** to every base.

They are an annoyance if you are using Fermat's Little Theorem to test for primality, because if your number just happens to be a Carmichael number, it will pass the test for any base—and still be composite.

Fortunately, the Carmichael number is quite rare. Those less than 100,000 are: 561, 1105, 1729, 2465, 2821, 6601, 8911, 10585, 15841, 29341, 41041, 46657, 52633, 62745, 63973, and 75361.

There are only 2,163 less than 25,000,000,000, and 105,212 less than 10^{15}, each with at most nine prime factors.

If n is a Carmichael number, then it is **squarefree**, the product of at least three distinct primes, and for every prime p divisor of n, $p - 1$ divides $n - 1$, and conversely. For example, 561 is the smallest Carmichael number and $561 = 3 \cdot 11 \cdot 17$, and 2, 10, and 16 all divide 560. The largest known Carmichael number with three prime factors was found by Harvey Dubner. It has 10,200 digits.

The smallest with four distinct factors is $41041 = 7 \cdot 11 \cdot 13 \cdot 41$, and 41040 is divisible by 6, 10, 12, and 40. The smallest with five distinct prime factors is 825265 and the smallest with six distinct prime factors is 321197185.

Carmichael conjectured in 1910 that there is an infinite number of Carmichael numbers. Alford, Granville, and Pomerance proved this in 1994 by showing how suitable **smooth numbers** could be multiplied together to fit the Carmichael definition.

Whether there is an infinity of Carmichael numbers with a given number of factors (at least three) is not known, however, nor whether Carmichael numbers can be found with an arbitrarily large number of factors.

Catalan's conjecture

Anyone might notice as a curiosity that 8 and 9 are 2^3 and 3^2, respectively, and that other small powers, such as 25 and 27, are not consecutive. Eugène Charles Catalan (1814–1894) conjectured in 1844 that 8 and 9 are indeed the only consecutive powers.

Levi ben Gerson (1288–1344) had shown that these are the only powers of 2 and 3 differing by 1, and **Euler** proved that 9 and 8 are the only square and cube differing by 1.

Robert Tijdeman proved in 1976 that the equation $x^p - y^q = 1$ has at most a finite number of solutions: if there is a solution, then p and q are less than a certain (unknown!) constant, C.

Computer checks show that if $x^p - y^q = \pm 1$, then p and q must exceed 10^7. It is also known that if $x^p - y^q = 1$, and if p and q are prime, then $p \mid y$ and $q \mid x$.

In 2000, Preda Mihailescu proved that if any solutions apart from 8 and 9 exist, then p and q must both be double **Wieferich primes**: $p^{(q-1)}$ must leave a remainder of 1 when divided by q^2, and $q^{(p-1)}$ must leave a remainder of 1 when divided by p^2. The only known examples are: 2 and 1093; 3 and 1006003; 5 and 1645333507; 83 and 4871; 911 and 318917; and 2903 and 18787.

It has also been proved by Hyyrö and Makowski that it is impossible to have three consecutive powers.

Catalan's Mersenne conjecture

When Lucas proved in 1876 that $2^{127} - 1$ is prime, Catalan noticed that $127 = 2^7 - 1$ and conjectured that this sequence, where M_p is the pth Mersenne number, contains only primes:

$$
\begin{aligned}
C_1 &= 2^2 - 1 = 3 &&= M_2 \\
C_2 &= 2^{C_1} - 1 = 2^3 - 1 &&= M_3 &= 7 \\
C_3 &= 2^{C_2} - 1 = 2^7 - 1 &&= M_7 &= 127 \\
C_4 &= 2^{C_3} - 1 = 2^{127} - 1 &&= M_{127} = \\
& \quad 170141183460469231731687303715884105727
\end{aligned}
$$

and so on . . .

Unfortunately, C_5 has more than 10^{38} digits and so cannot be tested directly, though Curt Noll has verified that C_5 has no prime divisor less than $5 \cdot 10^{50}$. Like so many conjectures of this kind, it is likely that a composite term appears quite soon.

See Mersenne numbers and Mersenne primes; strong law of small numbers

Champernowne's constant

David Champernowne (1912–2000) discussed *Champernowne's constant* in 1933: 0.12345678910111213 . . . It is transcendental (Mahler 1961: *MathWorld*) and normal in base 10, meaning that each digit 0 to 9 occurs one-tenth of the time, each pair of digits from 00 to 99 occurs one-hundredth of the time, and so on.

champion numbers

Conway and Odlyzko call the difference $p_{n-1} - p_n$ a "champion for x," denoted by $C(x)$, if it happens that it occurs most frequently for all the consecutive primes less than x.

$C(x)$ seems to take only the value 4, plus the values 2, 6, 30, 210, 2310, . . . which are the **primorials**, the result of multiplying the consecutive primes together. Is this true? Marek Wolf, Odlyzko, and Rubinstein say yes. (Rivera, Conjecture 10)

Chinese remainder theorem

Sun Tsu Suan-ching (fourth century AD) posed this problem: "There are certain things whose number is unknown. Divided by 3, the remainder is 2; by 5 the remainder is 3; and by 7 the remainder is 2. What will be the number?" The solution is 23. (Wells 1992, 23)

This is an example of the Chinese remainder theorem, which says that if you know the remainders when N is divided by n numbers, *which are coprime in pairs*, then you can find a unique smallest value of N, and an infinity of other solutions, by adding any integral multiple of the product of the n numbers (or subtracting if you are satisfied with negative solutions).

In Sun Tsu Suan-ching's puzzle, $3 \cdot 5 \cdot 7 = 105$, so the solutions are 23, 23 + 105, 23 + 210, 23 + 315, and so on, and $23 - 105 = -82$ is the smallest negative solution.

The Chinese remainder theorem can also be expressed in terms of **congruences**: if

$$x \equiv r_1 \ (\text{mod} \ m_1)$$
$$x \equiv r_2 \ (\text{mod} \ m_2)$$
$$x \equiv r_3 \ (\text{mod} \ m_3)$$
$$\ldots$$
$$x \equiv r_n \ (\text{mod} \ m_n)$$

then there is a unique solution, X, for x lying between 0 and $m_1 m_2 \ldots m_n$, and the general solution is congruent to X (mod $m_1 m_2 \ldots m_n$).

One use of the Chinese remainder theorem is to do arithmetic on large numbers by choosing a set of moduli m_1, m_2, ... m_n and then treating each number as a set of remainders, r_1, r_2, r_3, ... r_n, rather than as a sequence of decimal or binary digits. Then you do the arithmetic on the remainders and recover the solution by using the Chinese remainder theorem.

cicadas and prime periods

Cicadas of the genus *Magicicada* appear once every 7, 13, or 17 years. Is it just a coincidence that these are prime numbers? Eric Goles, Oliver Schulz, and Mario Markus have found evolutionary predator-prey models that have prime periods—which they then used to generate large prime numbers. (Sugden 2001, 177)

circle, prime

Is it always possible to arrange the numbers from 1 to $2n$ in a circle so that each adjacent pair sums to a prime?

Antonio Filz calls such an arrangement a prime circle. For example, these are the essentially unique prime circles for $n = 1$, 2, and 3:

```
                1 4              1 6
    1 2                       4       5
                2 3              3 2
```

There are two prime circles for $n = 4$ and forty-eight for $n = 5$. It is not known if there are prime circles for all values of n.

circular prime

A prime is circular if all the cyclic permutations of its digits are prime. These primes and their cyclic permutations are circular, in base 10:

2, 3, 5, 7, R_2, 13, 17, 37, 79, 113, 197, 199, 337, 1193, 3779, 11939, 19937, 193939, 199933, R_{19}, R_{23}, R_{317}, R_{1031}

where R_n stands for the nth **repunit prime**.

Walter Schneider has checked that there are no more up to 10^{22}.
  See permutable primes

Clay prizes, the

In Paris at the Collège de France on May 24, 2000, almost exactly one hundred years since David **Hilbert's 23 problems** were presented to the world, seven new "Millennium Prize Problems" were announced, for which the Clay Mathematics Institute of Cambridge, Massachusetts, is offering prizes of $1 million to the first solver of each problem.

All the Clay problems are, of course, extremely difficult, and have resisted the attempts of mathematicians for many years, but one problem is outstanding: the only one from Hilbert's original 23 that appears in the Clay list is the **Riemann hypothesis**.

As a protection against the naive or frivolous claims that such a large prize is sure to provoke, solvers must not send their claims directly to the Clay Institute but must get them published in a math-ematics journal of worldwide repute and the claimed solution must then be accepted by the mathematics community. Two years is allowed for this process. If the solution survives scrutiny, only then will it be considered by the Scientific Advisory Board of the Clay Mathematics Institute.

The procedure is slightly different if the claim is for a counterexam-ple, so if you think you have found a zero of the Riemann zeta func-tion that does not have real part ½, see the Clay Institute Web site for what to do. However, since Andrew Odlyzko has calculated a million zeros near zero number 10^{20} and *ten billion* zeros near zero number 10^{22}, and the ZetaGrid **distributed computing** network is calculating more than a billion zeros a day, you'd better get your skates on!

There is a second Clay prize challenge that is relevant to the primes: the *P* versus *NP* problem. It is currently very hard to factorize large numbers but quick and easy to check the factorization once it is found. Is there really no way to factorize large numbers quickly?

If you do discover a method, you might plausibly earn far more than the Clay Institute's $1 million by selling your discovery to commercial organizations—or governments—who use numbers that are the product of two large primes for **public key encryption** and would be very interested to hear that their secure communications can be broken using your method!

See AKS algorithm; distributed computing; factorization; public key encryption; Riemann hypothesis

compositorial

The product of all the composite numbers less than or equal to n is $n!$ (*n*-**factorial**) divided by the product of the primes less than or equal to n, or n-**primorial**, denoted by $n\#$. Iago Camboa has suggested calling this *n-compositorial*. (Caldwell, *Prime Pages*) Just as $n!$ and $n\#$ have many factors, so does $n!/n\#$, so $n!/n\# \pm 1$ is relatively likely to be prime.

$n!/n\# + 1$ is prime for $n = 1, 2, 3, 4, 5, 8, 14, 20, 26, 34, 56, \ldots$
$n!/n\# - 1$ is prime for $n = 4, 5, 6, 7, 8, 14, 16, 17, 21, 34, 39, \ldots$

See also factorial; primorial

concatenation of primes

The concatenation of the primes gives the sequence:

$$2, 23, 235, 2357, 235711, \ldots$$

The nth term is prime for $n = 1, 2, 4, 128, 174, 342, 435, 1429, \ldots$ with no others less than 7837. (Weisstein, 2001)

The Copeland-Erdös constant is the decimal $0.23571113171923\ldots$ Copeland and Erdös (1946) showed that it is normal in base 10. It is also irrational, as is the decimal number $.0110101000101000101\ldots$ in which the nth digit is 1 if n is prime and 0 otherwise.

See Champernowne's constant

conjectures

The theory of numbers, more than any other branch of pure mathematics, has begun by being an empirical science. Its most famous theorems have all been conjectured, sometimes a hundred years or more before they have been proved; and they have been suggested by the evidence of a mass of computation.

—*G. H. Hardy (1920, 651)*

One of the delights of prime numbers is that their combination of strict definition with apparent irregularity amounting almost to randomness invites mathematicians both professional and amateur to propose more and more problems and conjectures, the best-known named after their proposers.

Many of these conjectures are extremely difficult to settle. The prime numbers are just *too* mysterious and difficult!

When a sudden illumination invades the mathematician's mind . . . it sometimes happens . . . that it will not stand the test of verification . . . it is to be observed that almost always this false idea, if it had been correct, would have flattered our natural instincts for mathematical elegance. (Henri Poincaré n.d.)

They are also too tempting. As Poincaré's comment suggests, it is oh so easy to spot an elegant pattern and assume that it goes on forever. How often it doesn't! How often we are disappointed!

As G. H. Hardy also noted, "Some branches of mathematics have the pleasant characteristic that what seems plausible at first sight is generally true. In [analytic prime number] theory anyone can make plausible conjectures, and they are almost always false." (Hardy 1915, 18)

The simplest conjectures are easy to make and may be easy to prove, though not as easily as in the joke about a physicist who notices that 3 is prime, 5 is prime, 7 is prime, 9 is not—but that's an experimental error!—11 is prime, 13 is prime . . . and so concludes that all odd number are prime!

The most important conjectures tend to be made by the most brilliant mathematicians who have looked extraordinarily deeply into the subject and whose intuition tells them that a certain "fact" is likely to be true, although they cannot prove it. Such deep conjectures have contributed enormously to the progress of mathematics.

Fermat's Last Theorem, labeled a "theorem" only because Fermat claimed to have proved it, was for centuries a plausible conjecture until it was finally proved by Andrew Wiles.

Today's most famous and hardest mathematical problem is by common consent the **Riemann hypothesis**, a conjecture about the distribution of the prime numbers.

Conjectures about prime numbers have another feature that can be both intriguing and infuriating. Because the primes are quite frequent among the "small" integers, there are many tempting conjectures that fail as soon as we get out a modern electronic calculator or a powerful computer.

Fermat's conjecture that $2^{2^n} + 1$ is always prime must have seemed very tempting to Fermat, but the very first value that he did not calculate, because it was so large, turned out to be composite! It is indeed easy to find functions that seem to produce many primes for small values of n, but which start to produce composites as n increases. How tempting to conjecture that we have discovered a **formula for primes** and how disappointing when the formula fails! Richard Guy has referred to this phenomenon as the **strong law of small numbers**.

On the other hand, some conjecture may seem very well founded indeed, because the first counterexample is so very large. In 1919, George Pólya, author of *Mathematical Discovery* and *Mathematics and Plausible Reasoning*, conjectured that the number of integers $\leq N$ with an odd number of prime factors is never less than the number of integers $\leq N$ with an even number of prime factors. (For the purposes of this conjecture, repeated factors are counted, $N = 1$ is counted as having no prime factor, and a prime is counted as having one prime factor.)

For nearly forty years this was believed to be true, though no one could prove it. Then in 1958 it was proved that it is false for infinitely many N, and in 1980 M. Tanaka showed that the smallest counterexample is when $N = 906,150,257$. (Haimo 1995)

You may well find yourself making a conjecture or two as you read this book: meanwhile, many well-known conjectures have their own entries or are listed in the index.

See induction; Riemann hypothesis; strong law of small numbers

consecutive integer sequence

The consecutive integer sequence goes: 1, 12, 123, 1234, 12345, . . . There are no primes among the first 13,500 terms. (Weisstein, *Math-World*)

consecutive numbers

Sylvester proved in 1892 that every product of n consecutive integers greater than n is divisible by a prime greater than n.

In fact, although the product of five consecutive integers $6 \cdot 7 \cdot 8 \cdot 9 \cdot 10$ is divisible by just one prime, 7, greater than 5, the product $200 \cdot 201 \cdot 202 \cdot 203 \cdot 204$ is divisible by five primes, 67, 101, 7, 29, and 17. This suggests that Sylvester's result is rather generous: indeed, the product of just two consecutive numbers is always divisible by a prime greater than N, if the product is large enough.

Since the triangular numbers have the formula $\frac{1}{2}n(n + 1)$, we could rephrase this to say that for *any* number N, all the triangular numbers from a certain point onward are divisible by a prime greater than N.

Below each of these nine composite numbers is one of its prime factors, and all these factors are different:

1802	1803	1804	1805	1806	1807	1808	1809	1810
53	601	41	19	43	139	113	67	181

In general, given n consecutive integers greater than $n^{n-1} + 1$, then each of them has a prime factor that divides none of the others. (Grimm 1969)

consecutive primes, sums of

In how many ways can a number, n, be written as the sum of *one or more* consecutive primes? If we call it $f(n)$, then $f(5) = 2$, because $5 = 5$ and $2 + 3$, and $f(41) = 3$, because $41 = 11 + 13 + 17 = 2 + 3 + 5 + 7 + 11 + 13$.

Leo Moser has proved that the average value of $f(n)$ from $n = 1$ to N is log 2 as N tends to infinity. (Guy 1981, C2)

See also Goldbach's conjecture

Conway's prime-producing machine

This cunning device consists of a row of fractions, which are labeled for easy reference:

$$\frac{17}{91} \quad \frac{78}{85} \quad \frac{19}{51} \quad \frac{23}{38} \quad \frac{29}{33} \quad \frac{77}{29} \quad \frac{95}{23} \quad \frac{77}{19} \quad \frac{1}{17} \quad \frac{11}{13} \quad \frac{13}{11} \quad \frac{15}{14} \quad \frac{15}{2} \quad \frac{55}{1}$$

$$A \quad B \quad D \quad H \quad E \quad F \quad I \quad R \quad P \quad S \quad T \quad L \quad M \quad N$$

You always start with the number 2. A step involves multiplying the current number by the earliest fraction in the machine that makes the answer a whole number.

The machine pauses whenever a power of 2 is reached, and the output is the exponent of that power of 2. Here is how it starts:

2	15	825	725	1925	2275	425	390	330	
	M	N	E	F	T	A	B	S	E

290	770	910	170	156	132	116	308	364	68	4
	F	T	A	B	S	E	F	T	A	P

It takes 19 steps for the number $4 = 2^2$ to appear, and 2 is the first prime! Continuing, after another 50 steps, $8 = 2^3$ appears, and 3 is the second prime. After another 211 steps, the next power of 2 is $32 = 2^5$, and 5 is the third prime. And so on. (Conway and Guy 1996, 130, 147) (Guy 1983)

cousin primes

Cousin primes are pairs differing by 4, so they are rather more distant than **twin primes** but less distant than **sexy primes**. There are fourteen pairs of twin primes less than 200, and also fourteen pairs of cousin primes: 3-7, 7-11, 13-17, 19-23, 37-41, 43-47, 67-71, 79-83, 97-101, 103-107, 109-113, 127-131, 163-167, and 193-197. There are twenty-six more pairs below 1000.

If the first of the **Hardy-Littlewood conjectures** is true, then the twin and cousin primes have the same density, as we move to infinity.

Based on the cousin primes up to 2^{42}, and omitting the exceptional initial pair, 3-7, because 3 is not of the form $6n + 1$, the series,

$$\tfrac{1}{7} + \tfrac{1}{11} + \tfrac{1}{13} + \tfrac{1}{17} + \tfrac{1}{19} + \tfrac{1}{23} + \tfrac{1}{37} + \tfrac{1}{41}$$

has the sum 1.1970449 . . . (Wolf 1996)

See sexy primes; twin primes

Cullen primes

Numbers of the form $C_n = n \cdot 2^n + 1$ are named after the Reverend J. Cullen, who noticed in 1905 that apart from $C_1 = 3$ and one other possible exception, they are all composite for $n = 1$ to 100. The exception was C_{53}, which was found by Cunningham to be divisible by 5591.

Although for low values of n, Cullen primes are rare, it has been conjectured that there is an infinite number of them.

The known Cullen primes occur when $n = 1, 141, 4713, 5795, 6611, 18496, 32292, 32469, 59656, 90825, 262419, 361275,$ and 481899.

Numbers of the form $n \cdot b^n + 1$, called generalized Cullen numbers, are also rarely prime.

When $b = 3$, $n \cdot 3^n + 1$ is prime for $n = 2, 8, 32, 54, 114, 414, 1400, 1850, 2848, 4874, 7268, 19290.$

The largest known Cullen prime is C_{481899} of 145,072 digits, discovered by Masakatu Morii in 1998.

See Woodall primes

Cunningham project

Lieutenant Colonel Allan Joseph Cunningham (1842–1928) retired from the British army in 1891 and devoted himself to number theory, especially the factorization of numbers of the form $a^n \pm b^n$.

In 1925 he got together with H. J. Woodall to publish a book of all that they had discovered about the factors of such numbers. D. H. Lehmer devoted much time to extending their results, in what is now called the Cunningham project. Of course, there were many numbers in their tables that at the time could not be factored with the equipment available. D. H. Lehmer and Emma Lehmer factored many of them. The last number in the original book was factored in 1992.

The project is continuing. The results of the Cunningham project were originally collected in J. Brillhart et al., *Factorizations of $b^n \pm 1$, b = 2, 3, 5, 6, 7, 10, 11, 12 up to high powers*, published by the American Mathematical Society in 1988. The results are now on the Web, where they are regularly updated at this site run by Stan Wagstaff: www.cerias.purdue.edu/homes/ssw/cun. This site includes a calculator that gives the known prime factors of the numbers $b^n \pm 1$ (where b = 2, 3, 4, 5, 6, 7, 8, 9, 10, 11, 12).

See GIMPS; Mersenne primes

Cunningham chains

If p and $2p + 1$ are both prime, then p is a **Sophie Germain prime**. A Cunningham chain is a chain of Sophie Germain primes, apart from the last prime in the chain. Two small examples: 2, 5, 11, 23, 47; and 89, 179, 359, 719, 1439, 2879.

These are sometimes called Cunningham chains of the first kind, to distinguish them from Cunningham chains of the second kind, in which all the terms are prime and each is double the previous prime, *less* 1: for example, 2, 3, 5; or 19, 37, 73.

If the strong prime **k-tuples conjecture** is true, then Cunningham chains can reach any length.

Tony Forbes found on December 5, 1997, the longest Cunningham chain of the first kind, with 14 terms, and the longest of the second kind, of length 16, starting with 3203000719597029781.

The largest chain of the first kind of length 3 starts at $115566729.2^{4319} - 1$, and the largest of the second kind of length 3 starts at $734257203.2^{5000} + 1$, both discovered by Warut Roonguthai. It is no coincidence that both feature numbers of the form $k \cdot 2^n \pm 1$. Although any odd number can be expressed in these forms, if $k < 2^n$ then the number can be tested for primality very efficiently. Since $2(k \cdot 2^n - 1) + 1 = k \cdot 2^{n+1} - 1$, a Cunningham chain may consist of a sequence of terms of this form. Similarly, $2(k \cdot 2^n + 1) - 1 = k \cdot 2^{n+1} + 1$, leading naturally to chains of the second kind.

decimals, recurring (periodic)

The fraction $1/7$ as a decimal is $0.142857\ 142857\ldots$ The number 142857 has many curious properties, often shared (more or less) with the periods of other primes. For example, splitting the period in two halves or three thirds:

$$142 + 857 = 999, \text{ and } 14 + 28 + 57 = 99$$

Square and split into halves and add: $142857^2 = 20{,}408{,}122{,}449$, and:

$$20{,}408 + 122{,}449 = 142857$$

Multiplying 142857 by any number from 2 to 6 results in a permutation of its digits:

$$142857 \cdot 2 = 285714$$
$$142857 \cdot 3 = 428571$$
$$142857 \cdot 4 = 571428$$
$$142857 \cdot 5 = 714285$$
$$142857 \cdot 6 = 857142$$

But,
$$142857 \cdot 7 = 999999$$

the period of 1/13

The fraction $1/13 = 0.076923\ 076923\ldots$ has period length $6 = (13 - 1)/2$. These properties match those of $1/7$:

$$076 + 923 = 999 \qquad 07 + 69 + 23 = 99$$

The second property needs only to be slightly adjusted:

$076923^2 = 005{,}917{,}147{,}929$, and adding the six-digit split halves we get,

$$005{,}917 + 147{,}929 = 153{,}846 = 2 \cdot 076923$$

The third property is also somewhat different. Multiplying 076923 by the numbers 2 to 12:

$076923 \cdot 2\ = 153846$		$076923 \cdot 3\ = 230769$
$076923 \cdot 4\ = 307692$		$076923 \cdot 5\ = 384615$
$076923 \cdot 6\ = 461538$		$076923 \cdot 7\ = 538461$
$076923 \cdot 8\ = 615384$		$076923 \cdot 9\ = 692307$
$076923 \cdot 10 = 769230$		$076923 \cdot 11 = 846153$
$076923 \cdot 12 = 923076$		

Five of the products are cyclic permutations of 076923 but the other six are cyclic permutations 153846.

We can put them into a simpler sequence if we start with 076923 and just double and double again, reducing modulo 13, so that $16 \equiv 3$ (mod 13) and $32 \equiv 6$ (mod 13) and so on. The sequence of multiples from 2 to 12 then reads: 2, 4, 8, 3, 6, 12, 11, 9, 5, 10, 7, and the products are cyclic permutations of 07923 and 153846 *alternately*.

The decimal reciprocals of composite numbers are, naturally, more complicated and less obviously elegant. For example, $1/21 = 0.047619$ and $047619 \cdot 21 = 999999$ and $04 + 76 + 19$ does equal 99— but $047 + 619 = 666$, not 999.

cyclic numbers

The periods of the reciprocals of the primes are also known as *cyclic numbers*. These are the periods of the other primes below 100, excluding 2, 3, 5, and 11:

period	$1/p$
13	076923
17	0588235294117647
19	052631578947368421
23	0434782608695652173913
29	0344827586206896551724137931
31	032258064516129 [the smallest with period $(p-1)/2$]
37	27
41	02439 [the smallest with period $(p-1)/8$]
43	023255813953488372093
47	0212765957446808510638297872340425531914893617
53	0188679245283 [the smallest with period $(p-1)/4$]
59	0169491525423728813559322033898305084745762711864406779661
61	01639344262295081967213114754098360655737704918032 7868852459
67	014925373134328358208955223880597
71	01408450704225352112676056338028169
73	01369863
79	0126582278481
83	01204819277108433734939759036144578313253

Continued on next page

89   0112359550561797752808988764044943820224719

97   0103092783505154639175257731958762886597938144329

If the period of a prime p is of length $p - 1$, it is a *full period prime* (also called a *reptend* or *long prime*). In all other cases the period length is a factor of $n - 1$, as Lambert noticed in 1769.

Artin's conjecture

Thirty-eight of the first hundred primes are full period, starting with 7, 17, 19, 23, 29, 47, 59, 61, . . . but no general method is known for deciding which primes are full period.

There is a connection here with primitive roots: if p is a prime less than 5, then the decimal expansion of $1/p$ has the maximum possible period of $p - 1$ in base 10 if and only if 10 is a **primitive root** modulo p.

Roughly, it seems by calculation that 37% of all primes in base 10 are full period. Emil Artin conjectured that the exact figure should be, *for any base at all*:

$$\frac{1 \times 5 \times 19 \times 41 \times 109 \times 155 \times 271}{2 \times 6 \times 20 \times 42 \times 110 \times 156 \times 272} \times \ldots = 0.3739558 \ldots = C$$

The fractions are $(p^2 - p - 1)/(p^2 - p)$ for each prime, 2, 3, 5, . . .

D. H. and Emma Lehmer then discovered experimentally that Artin's conjecture is not quite right. It requires a correcting factor that depends on the base. In base 7, for example, the proportion is conjectured to be $42C/41$.

the repunit connection

There is a simple connection between reciprocal prime periods and the **repunits**, numbers whose digits are all 1, which we can illustrate with the cases of 1/7 and 1/13:

	1/7	= 0.142857 142857 . . .
and	7 · 1/7	= 1 = 0.999999 999999 . . .
so,	7 · 142857	= 999999 = 9 · 111111

	1/13	= 0.076923 076923 . . .
Similarly,	1/13	= 0.076923 076923 . . .
and	13 · 1/13	= 1 = 0.999999 999999 . . .
so,	13 · 076923	= 999999 = 9 · 111111

In other words, all the integers whose reciprocals have period 6 must divide $999999 = 3^3 \cdot 7 \cdot 11 \cdot 13 \cdot 37$. As it happens, 3, 11, and 37 have periods less than 6, so only 1/7 and 1/13 actually have period 6. Similarly, since $111,111,111 = 3 \cdot 3 \cdot 37 \cdot 333667$, the only prime with reciprocal period length 9 is 333667. These primes have unique period lengths: 3, 11, 37, 101, 333667, 9091, 9901, 909091, R_{19}, R_{23}, 99990001, 999999000001, 909090909090909091, . . . (Sloane M2890)

magic squares

The periods of 1/7, 2/7, . . . 6/7 form an imperfect magic square, because although the rows and columns have the same sum, the diagonals do not:

$$
\begin{array}{cccccc}
1 & 4 & 2 & 8 & 5 & 7 \\
2 & 8 & 5 & 7 & 1 & 4 \\
4 & 2 & 8 & 5 & 7 & 1 \\
5 & 7 & 1 & 4 & 2 & 8 \\
7 & 1 & 4 & 2 & 8 & 5 \\
8 & 5 & 7 & 1 & 4 & 2 \\
\end{array}
$$

The decimal periods of 1/19, 2/19, . . . , 18/19, however, do form a true magic square.

1/19	=	0	5	2	6	3	1	5	7	8	9	4	7	3	6	8	4	2	1
2/19	=	1	0	5	2	6	3	1	5	7	8	9	4	7	3	6	8	4	2
3/19	=	1	5	7	8	9	4	7	3	6	8	4	2	1	0	5	2	6	3
4/19	=	2	1	0	5	2	6	3	1	5	7	8	9	4	7	3	6	8	4
5/19	=	2	6	3	1	5	7	8	9	4	7	3	6	8	4	2	1	0	5
6/19	=	3	1	5	7	8	9	4	7	3	6	8	4	2	1	0	5	2	6
7/19	=	3	6	8	4	2	1	0	5	2	6	3	1	5	7	8	9	4	7
8/19	=	4	2	1	0	5	2	6	3	1	5	7	8	9	4	7	3	6	8
9/19	=	4	7	3	6	8	4	2	1	0	5	2	6	3	1	5	7	8	9
10/19	=	5	2	6	3	1	5	7	8	9	4	7	3	6	8	4	2	1	0
11/19	=	5	7	8	9	4	7	3	6	8	4	2	1	0	5	2	6	3	1
12/19	=	6	3	1	5	7	8	9	4	7	3	6	8	4	2	1	0	5	2
13/19	=	6	8	4	2	1	0	5	2	6	3	1	5	7	8	9	4	7	3
14/19	=	7	3	6	8	4	2	1	0	5	2	6	3	1	5	7	8	9	4
15/19	=	7	8	9	4	7	3	6	8	4	2	1	0	5	2	6	3	1	5
16/19	=	8	4	2	1	0	5	2	6	3	1	5	7	8	9	4	7	3	6
17/19	=	8	9	4	7	3	6	8	4	2	1	0	5	2	6	3	1	5	7
18/19	=	9	4	7	3	6	8	4	2	1	0	5	2	6	3	1	5	7	8

The row, column, and diagonal sums are all 81. Notice the patterns in the first and last columns. (Caldwell, *Prime Pages*)

deficient number

A number is deficient if the sum of its *proper* divisors, meaning all its divisors except the number itself is less than the number: so all prime numbers are deficient.

The number 8 is deficient, because $1 + 2 + 4 = 7$. In fact, all powers of 2 are deficient just because the proper factors of 2^n are 1, 2, 4, 8, ..., 2^{n-1} and $1 + 2 + 4 + ... + 2^{n-1} = 2^n - 1$. The powers of primes greater than 2 are even more deficient.

The opposite of a deficient number is an **abundant number**.

See abundant number; perfect number

deletable and truncatable primes

Chris Caldwell defines a *deletable prime* to be one that remains prime as the digits are deleted in some chosen order. This is his example: 410256793, 41256793, 4125673, 415673, 45673, 4567, 467, 67, 7. It is not known whether there is an infinity of such primes.

truncatable primes

A right-truncatable number is prime and remains prime as the digits are removed from the right. It therefore contains no zero digit, and the digits 2 and 5 can only be the leftmost digit.

There are eighty-three right-truncatable primes in base 10, starting, 2, 3, 5, 7, 23, 29, 31, 37, 53, 59, 71, 73, 79, 233, 239, 293, 311, 313, 317, 373, ...

There is an infinity of left-truncatable primes if zeros are allowed; for example, 1087. If zeros are disallowed, there are 4,260 left-truncatable primes in base 10, starting 2, 3, 5, 7, 13, 17, 23, 37, 43, 47, 53, 67, 73, 83, 97, 113, 137, 167, 173, ...

Demlo numbers

The original Demlo numbers are the palindromes 1, 121, 12321, 1234321, 123454321, ..., 12345678987654321. The problem is then, how to continue the sequence?—and the usual solution is to notice that the Demlo numbers are the squares of the first few **repunits**:

$$11^2 \quad = 121$$
$$111^2 \quad = 12321$$
$$1111^2 = 1234321$$

$$\cdots$$

$$R_9{}^2 = 111111111^2 = 12345678987654321$$

This suggests that Demlo-10 should be defined as,

$$R_{10}{}^2 = 1111111111^2 = 1234567900987654321$$

and so on.

descriptive primes

In a descriptive (or self-descriptive or Look and Say) sequence, each term *describes* the previous term. For example,

2	12	1112	3112
	one 2	one 1, one 2	three 1s, one 2 . . .

Starting with 1, the sequence continues 1, 11, 21, 1211, 111221, 312211, 13112221, . . . The first two primes are 11 and 312211.

Are there descriptive sequences whose terms are all prime? G. L. Honaker Jr. found this one: 373, 131713, 111311171113, 311331173113. Unfortunately, the next term is composite. Carlos Rivera, Mike Keith, and Walter Schneider have subsequently found six-term sequences, and Schneider has found a seven-term sequence starting with 19,972,667,609. (Schneider 2003) (Rivera, Puzzle 36)

Dickson's conjecture

Leonard Eugene Dickson (1874–1954) is best known today for his extraordinarily detailed three-volume *History of the Theory of Numbers*, whose first volume is on *Divisibility and Primality*. He conjectured in 1904 that if we have a sequence of linear expressions with integer coefficients, with all the a_i positive,

$$a_1n + b_1, \ a_2n + b_2, \ a_3n + b_3, \ \ldots, \ a_kn + b_k$$

then there is infinite number of values of n for which every one of these expressions will be prime simultaneously (apart from cases where there is a number which divides at least one the expressions for every value of n). The qualification is necessary to exclude cases such as,

$$2n + 3 \quad 2n + 5 \quad 2n + 7$$

one of which must always be divisible by 3: the first if $3 \mid n$, the second if $3 \mid n + 1$, the third if $3 \mid n - 1$.

Dickson's general conjecture includes many others: the **twin primes** conjecture is the case for n and $n + 2$. If it is true for n and $2n + 1$, then there is an infinity of **Sophie Germain** primes. It also implies, if true, the prime **k-tuples conjecture**; that there are infinitely many triples of consecutive **semiprimes**; and that there is an infinity of composite **Mersenne numbers**.

See hypothesis H

digit properties

The largest prime using the digits 1 to 9 is 98765431. If 0 can be used as well, it is 987654103.

- 8757193191 is the largest prime such that the first n digits are divisible by the nth prime, for $n = 1$ to 10. (Mike Keith: Caldwell, *Prime Pages*)
- 113 is the smallest three-digit prime such that all rearrangements of its digits are also prime. The others are 337 and 199. The **repunits** R_{19} and R_{23} are the next numbers with this property.
- 200 is the least number that cannot become a prime by changing one digit. With 202, 204, 206, and 208 it forms an arithmetic progression of numbers with the same property.

See deletable and truncatable primes

Diophantus (c. AD 200; d. 284)

Diophantus was one of the great late Greek mathematicians. Six books of his *Arithmetica*, out of thirteen, are extant. He also wrote *On Polygonal Numbers*.

The *Arithmetica* is a collection of solved problems, including this one: which numbers can be the hypotenuse of a right-angled triangle?

According to Pythagoras's theorem, in a right-angled triangle $a^2 + b^2 = c^2$. The simplest example is $3^2 + 4^2 = 5^2$ and the next simplest is $5^2 + 12^2 = 13^2$. The general formula for solutions to Pythagoras's equation is:

$$a = m^2 - n^2 \qquad b = 2mn \qquad c = m^2 + n^2$$

If m and n have no common factor, then a, b, and c will be coprime also.

From the specific examples that Diophantus chose in solving his problems, critics have concluded that he knew that any prime number of the form $4n + 1$ is a possible hypotenuse, meaning that it is of the form $x^2 + y^2$.

He also knew that no number of the form $4n + 3$ is the sum of two squares, and numbers of the form $8n + 7$ are not the sum of three squares, though any number is the sum of up to four squares.

His *Arithmetica* includes the problem of representing numbers as the sum of two squares, which were rational rather than integral, for example:

$$x^2 + y^2 = 13, \text{ with } x^2 \text{ and } y^2 \text{ greater than } 6$$

He finds the solution is $(257/101)^2$ and $(258/101)^2$, which is equivalent to $257^2 + 258^2 = 13 \cdot 101^2$, an impressive feat! (Today, ironically, "Diophantine" means soluble in integers only.)

He also used relationships such as 8 times a triangular number + 1 = a square number; $m^2 + n^2 \pm 2mn$ is a square; the sum of two cubes is also the difference of two cubes (rational rather than integral); and $(m^2 - n^2)^2 + (2mn)^2 = (m^2 + n^2)^2$. This last represents the sides of a right-angled Pythagorean triangle.

He also used the formulae,

$$(a^2 + b^2)(c^2 + d^2) = (ac + bd)^2 + (ad - bc)^2$$
and
$$(a^2 + b^2)(c^2 + d^2) = (ac - bd)^2 + (ad + bc)^2$$

which he used to find four right-angled triangles with the same hypotenuse. For example,

$$(1^2 + 2^2)(3^2 + 1^2) = (1 \cdot 3 + 2 \cdot 1)^2 + (1 \cdot 1 - 2 \cdot 3)^2 = 5^2 + 5^2 = 50$$

and so

$$(1^2 + 2^2)(3^2 + 1^2) = (1 \cdot 3 - 2 \cdot 1)^2 + (1 \cdot 1 + 2 \cdot 3)^2 = 1^2 + 7^2 = 50$$

These algebraic identities are more than a curiosity. They make a crucial and deep link between "being a sum of two squares" and factors and products, and they suggest questions such as:

A. Can the factors of *any composite* number that is the sum of two squares be written themselves as the sum of two squares?

B. Can a prime number that has no proper factors be written as the sum of two squares?

C. In how many ways can a number with three factors, each of which is the sum of two squares, be written as the sum of two squares?

The works of Diophantus were published by, among others, Bachet de Méziriac in 1621, whose book was studied by **Fermat**.

The questions proposed and answered by Diophantus provide a link from Pythagoras and the earliest Greek mathematics to Fermat and other mathematicians of the seventeenth and eighteenth centuries who made these the first deep problems in modern number theory.

See also Euclid; Fermat

Dirichlet's theorem and primes in arithmetic series

Gustav Peter Lejeune Dirichlet (1805–1859) was a prodigy who as a young man carried with him on his travels his dog-eared copy of Gauss's *Disquisitiones Arithmeticae*. At the age of only twenty, he presented to the French Academy of Sciences a paper on equations of the form,

$$x^5 + y^5 = A \cdot z^5$$

A few weeks later he proved that **Fermat's Last Theorem** when $n = 5$ has no solution.

Euclid proved that there is an infinity of primes among the positive integers, but how many are there in arithmetic progressions such as

	1	5	9	13	17	21	25	29	...
or	2	7	12	17	22	27	32	37	...?

Dirichlet proved in 1837, a conjecture made by Gauss: if a and b are coprime positive integers, then the arithmetic progression a, $a + b$, $a + 2b$, $a + 3b$, ... contains infinitely many primes. He did so by proving that if p is a prime of the form $an + b$, with a and b coprime, then the sum of all the primes p of this form less than x is approximately,

$$\frac{1}{\phi(a)} \cdot \log \log x$$

as x tends to infinity. In other words, it increases without limit, albeit very slowly, and so the primes of that form cannot be finite in number. He also proved that the number of primes in the sequence less than n tends to

$$\frac{n}{\phi(a) \log n}$$

as n increases.

This proof represented the birth of *analytic* number theory, which uses calculus to draw conclusions about the integers.

Where does the first prime occur in an arithmetic sequence? See **Linnik's constant** for a partial answer.

primes in polynomials

Dirichlet also proved that if a, $2b$, and c have no common prime factor, then the quadratic expression $ax^2 + 2bxy + cy^2$ takes an infinity of prime values.

See Hardy; Hardy-Littlewood conjectures

distributed computing

How much of the power of your computer do you actually use? Almost certainly, very little. One way to use the "wasted" power is to get together with other computer users, take a very difficult problem that can be split into many smaller problems—and *distributed* via the Internet—and tackle one small problem each. That's distributed computing!

In round orders of magnitude, a typical personal computer will soon execute 100 million instructions per second; it will have 100 megabytes of memory and a gigabyte of disk storage; it will consume 100 watts of electricity and cost $1,000; 100 million of these machines will be attached to the Internet. Multiply it out: 10 quadrillion instructions per second, 10 billion megabytes of memory, 100 million gigabytes of disk storage, 10 gigawatts of electric-power demand, a price tag of $100 billion. It's probably worth rewriting your software to gain access to such a machine. (Hayes 1998)

You might be able to crack a very difficult code, analyze geological data, or calculate the shapes of molecules—or you could attempt to communicate with an alien civilization by joining the SETI@home project searching for signs of life in signals from the radio telescope at the Arecibo Observatory in Puerto Rico: so far more than 70,000 enthusiasts have signed up.

The biggest number-theoretic opportunities include factoring very large numbers and finding record-breaking prime numbers. Arjen Lenstra and Mark Manasse organized the first Internet factoring project in 1988. Soon their volunteers were factoring 100-digit numbers with ease, and in 1993 a team of six hundred successfully factored RSA-129 for a prize of $100, and rather more glory.

Among the many projects now running, the distributed computing search for **Fermat number** divisors has its Web address www.fermatsearch.org/particip. Twenty-eight researchers from Brazil, Finland, Germany, Greece, Iran, Italy, Japan, Spain, Sweden, Russia, and the USA are taking part, including Tony Forbes, author of the MFAC program that was used to find a factor of F_{31}.

Tony Forbes is also organizing an international search for a factor of $MM61 = 2^{M_{61}} - 1$. This is a **Mersenne number** with a Mersenne prime exponent, and the smallest whose primality has not been decided.

The **Sierpinski numbers** offer another challenge. In March 2002 there were only seventeen candidate values of k left to check: 4847, 5359, 10223, 19249, 21181, 22699, 24737, 27653, 28433, 33661, 44131, 46157, 54767, 55459, 65567, 67607, and 69109. So Louis K. Helm and David A. Norris started their "Seventeen or Bust" project. By the end of 2002 they had eliminated five candidates. On November 27, 2002, Stephen Gibson reported that $46157 \cdot 2^n + 1$ is a prime when $n = 698207$. On Dec. 2, 2002, James Burt discovered that $65567 \cdot 2^n + 1$ is a prime when $n = 1013803$.

Three days later (!) the computer of an anonymous participant showed that $44131 \cdot 2^n + 1$ is a prime when $n = 995972$. On December 7, Sean DiMichele reported that $69109 \cdot 2^n + 1$ is a prime when $n = 1157446$, and Peter Coels has discovered that $54767 \cdot 2^n + 1$ is a prime when $n = 1337827$. This prime has 402,569 digits, making it the seventh largest known prime.

The goal of ZetaGrid, organized by Sebastian Wedeniwski of IBM, is to calculate zeros of the **Riemann** zeta function. So far they have calculated nearly 400 billion. Currently, ZetaGrid links more than 10,000 workstations, has a performance rate of about 5649 GFLOPS, and calculates more than 1 billion zeta function zeros every day.

Not surprisingly, distributed computing can be highly competitive! According to the ZetaGrid Web site, the Top Team for the "last 7 days" on Saturday, November 6, 2004, was "Debian Linux Users Everywhere," with two active members, and thirty-six computers that calculated 321,992,600 zeros.

If mere glory doesn't grab you, then there are prizes. ZetaGrid is offering the following four prizes in accordance with [these] rules:

- $10 (USD) will be awarded to the first person who discovers the first two zeros that have a distance less than 10^{-6}, using the software provided by ZetaGrid.
- $100 (USD) will be awarded to the first person who discovers the first two zeros that have a distance less than 10^{-7}, using the software provided by ZetaGrid.
- $1,000 (USD) will be awarded to the first person who discovers a nontrivial zero that is not on the critical line, using the software provided by ZetaGrid, but only if this constitutes the first disproof of the **Riemann hypothesis** by any method.
- Up to $1 million (USD) will be awarded to the first 100 top producers of ZetaGrid if Sebastian Wedeniwski wins the $1 million prize for the proof of the Riemann hypothesis from Clay Mathematics Institute by using the results of the statistical summaries of ZetaGrid.

See Electronic Frontier Foundation; Generalized Fermat numbers; GIMPS; RSA Factoring Challenge

divisibility tests

An integer in base 10 is divisible by 2 if the last digit is even; by 3 if 3 divides the sum of the digits; by 9 if 9 divides the sum of the digits; and by 5 if the last digit is 5 or 0.

There are many tests for divisibility by 7, none of them very short. Here are two:

1. Multiply the left-hand digit by 3 and add to the next digit. Reduce the answer modulo 7 (meaning, take only the remainder when it is divided by 7). Repeat. If the final answer is divisible by 7, so was the original number.

Example: 6475: $6 \cdot 3 + 4 = 22$, which reduces to 1; then $3 \cdot 1 + 7 = 10$, reducing to 3; then $3 \cdot 3 + 5 = 14$, which is a multiple of 7.

2. Double the last digit and subtract it from the remaining number. Repeat. If the final result is 0 or ± 7, the original number is divisible by 7.

Example: $1106 \rightarrow 110 - 12 = 98 \rightarrow 9 - 16 = -7$
Example: $37989 \rightarrow 3798 - 18 = 3780 \rightarrow 378 \rightarrow 37 - 16 = 21$

So both numbers are divisible by 7.

There are simple tests for some other numbers. For example, $100a + b$ is divisible by 19 if and only if $a + 4b$ is, because $4(100a + b) = 400a + 4b \equiv a + 4b \pmod{19}$. If you test a number that cycles, such as,

$$1064 \rightarrow 10 + 256 = 266 \rightarrow 2 + 4 \cdot 66 = 266 \ldots$$

repeating, endlessly, then you're okay; the original number is divisible by 19 ($1064 = 19 \cdot 56$).

divisors (factors)

Man has long been aware that some numbers are more readily divided into parts than others, and that this can be a source of convenience. The Babylonians divided the sky and the circle into 360 degrees—suggested, plausibly, by the 365¼ days in the year—the day into 12 hours, the hour into 60 minutes, and they counted in 60s.

Plato in Book V of his *Laws* claims, "There is no difficulty in perceiving that the twelve parts admit of the greatest number of divisions of that which they include, or in seeing the other numbers which are consequent upon them," and he went on to argue that in his ideal republic,

> The number of our citizens shall be 5040, this will be a convenient number. . . . Every legislator ought to know so much arithmetic as to be able to tell what number is most likely to be useful to all cities; and we are going to take that number which contains the greatest and most regular and unbroken series of divisions. The whole of number has every possible division, and the number 5040 can be divided by exactly fifty-nine divisors [sixty including itself], and ten of these proceed without interval from one to ten; this will furnish numbers for war and peace, and for all contracts and dealing, including taxes and divisions of the land.

When the ancient Greeks first considered abundant, perfect, and deficient numbers, they tended to think of divisors as being less than the number, so the number itself was excluded: 28 was perfect because $28 = 1 + 2 + 4 + 7 + 14$.

However, there is a very good and typically logical reason for including n when calculating $d(n)$. The function $d(n)$ is *multiplicative,* meaning that provided a and b are coprime, then $d(ab) = d(a)d(b)$. So $d(n)$ can easily be calculated for any number whose prime factors are known. If we exclude the number itself, so we are considering $d(n) - n$, this is not multiplicative, and all our calculations become more complicated.

how many divisors? how big is *d(n)*?

A prime number, p, has two divisors, 1 and p. Powers of 2, 2^n, have $n + 1$ divisors, $1, 2, 4, \ldots, 2^n$, and the product of three distinct primes, pqr, has eight factors: $1, p, q, r, qr, rp, pq, pqr$.

In general, if n is written as the product of prime factors: $n = p^a q^b r^c \ldots$ then the number of divisors, $d(n) = (a + 1)(b + 1)(c + 1) \ldots$

Since half of all integers are divisible by 2, and a third by 3, and so on, we might think that most integers have quite a few divisors. This is false. The opposite, correct argument is that half of all integers are even, 1 in 6 is divisible by 2 and 3, 1 in 12 by 2, 3, and 4, only 1 in 30 is divisible by 2, 3, and 5, and only 1 in 210 is divisible by 2, 3, 5, and 7. So numbers with even a handful of *small* divisors are infrequent.

In fact, G. H. Hardy proved that a "typical" number, n, has about log log n divisors. Only a tiny proportion has many more divisors than this. The typical integer round about 10^8 has just three prime factors, and you have to go up to about 10^{70} to get an average of five prime factors.

The sum of the number of divisors of all the numbers up to n, $d(1) + d(2) + d(3) + \ldots + d(n)$ is approximately $n \log n$. More precisely it equals $n(\log n + 2\gamma - 1)$ as n tends to infinity, where γ is **Euler's constant**.

record numbers of divisors

We reach a record with $d(n)$ whenever we get to the smallest number with a given number of divisors. This, apart from 1, is always of the form $2^a \cdot 3^b \cdot 5^c \cdot 7^d \ldots$ where $a \geq b \geq c \geq d \ldots$ The sequence of such numbers starts,

n	$2^a \cdot 3^b \cdot 5^c \ldots$	$d(n)$
2	2^1	2
4	2^2	3
6	$2 \cdot 3$	4
16	2^4	5
12	$2^2 \cdot 3$	6
64	2^6	7
24	$2^3 \cdot 3$	8
36	$2^2 \cdot 3^2$	9
48	$2^4 \cdot 3$	10
1024	2^{10}	11
60	$2^2 \cdot 3 \cdot 5$	12
4096	2^{12}	13
192	$2^6 \cdot 3$	14
144	$2^4 \cdot 3^2$	15
120	$2^3 \cdot 3 \cdot 5$	16

curiosities of d(n)

- The first pair of consecutive numbers with the same number of divisors is 2 and 3, with two each. The subsequent pairs start,

n	14, 15	21, 22	26, 27	33, 34, 35	38, 39	44, 45
$d(n)$	4	4	4	4	4	6

Larger examples are 242 to 245, all with $d(n) = 6$, and 11605 to 11609, with $d(n) = 8$. (Rivera: Caldwell, *Prime Pages*)
- The product $nd(n)$ has equal values for each of the triplet 168, 192, and 224. There are three smaller pairs for which $nd(n)$ has the same values: 18 and 27; 24 and 32; 56 and 64. (Guy 1981, 68)

- The product of the harmonic mean and arithmetic mean of the divisors of a number is the number itself. For example, 20 has divisors 1, 2, 4, 5, 10, 20. The harmonic mean is the reciprocal of $(1/1 + 1/2 + 1/4 + 1/5 + 1/10 + 1/20)/6 = 2\frac{1}{10}/6 = 7/20$. So the harmonic mean is 20/7. The arithmetic mean is $42/6 = 7$.
- Call the maximum power of a prime that divides N a principal divisor of N. So 3 and 4 are principal divisors of 12. Then any odd integer N greater than 15 that is not a prime power is greater than twice the sum of its principal divisors. (Alspach 2004)

divisors and congruences

The simplest conclusions about divisibility come from looking at remainders. For example, the remainders when 2^n is divided by 7 are only ever 1, 2, or 4. In other words,

$$2^n \equiv 1, 2, \text{ or } 4 \pmod 7$$

It follows that $2^n + k$ can only be divisible by 7 when $k = 6, 5,$ or 3. For each value of n, only one of these will apply. For example, $2^7 = 128$ and 7 divides $128 + 5 = 133$.

Similarly, the first seven values of $x^2 + 3x + 5$ are:

x	1	2	3	4	5	6	7	
$x^2 + 3x + 5$	9	15	23	33	45	59	75	
	2	1	2	5	3	3	5	(mod 7)

So $x^2 + 3x + 5 \equiv 1, 2, 3,$ or 5 (mod 7) and its values are never divisible by 7.

the sum of divisors function

The sum of all the $d(n)$ divisors of n is written $\sigma(n)$. Like $d(n)$, $\sigma(n)$ is multiplicative: if p and q are coprime, then $\sigma(pq) = \sigma(p)\sigma(q)$.

If p is prime, then $\sigma(p) = p + 1$, so if $n = p^a q^b r^c \dots$ then,

$$\sigma(n) = (p^{a+1} - 1)(q^{b+1} - 1)(r^{c+1} - 1) \dots /(p - 1)(q - 1)(r - 1) \dots$$

The sequence of integers that are never values of $\sigma(n)$ starts:

2, 5, 9, 10, 11, 16, 17, 19, 21, 22, 23, 25, 26, 27, 29, 33, 34, 35, 37, 41, 43, 45, 46, 47, 49, 50, 51, 52, 53, . . .

(Sloane A007369)

$\sigma(n)$ is odd if and only if n is a square or double a square.

the size of σ(n)

For all $n > 1$,

$$\frac{\sigma(n)}{n\sqrt{n}} < 1$$

and for all n except $n = 2, 3, 4, 6, 8,$ and $12,$

$$\frac{\sigma(n)}{n\sqrt{n}} < 6/\pi^2$$

(Annapurna 1938)

As n tends to infinity, $\sigma(n)$ is of the order of $n \log \log n$. The upper limit of $\sigma(n)/n \log \log n$ as n tends to infinity is e^{γ}, where γ is **Euler's constant**.

The sum of the divisors $\sigma(n)$ does not jump around as much as $d(n)$, and the sum $\sigma(1) + \sigma(2) + \sigma(3) + \sigma(4) + \ldots + \sigma(n)$ is even smoother: it is approximately equal to $\pi^2 n^2/12$ plus a factor that is proportional to $n \log n$.

For example, the sum $\sigma(1) + \sigma(2) + \sigma(3) + \ldots + \sigma(1000) = 823081$ and $\pi^2 1000^2/12 = 822467$ to the nearest integer. (Lehmer 1940)

a recursive formula

Euler discovered an extraordinary and exquisite recursive formula for calculating $\sigma(n)$:

$$\sigma(n) = \sigma(n-1) + \sigma(n-2) - \sigma(n-5) - \sigma(n-7) + \sigma(n-12) + \sigma(n-15) - \sigma(n-22) - \sigma(n-26) + \sigma(n-35) + \sigma(n-40) \ldots$$

You stop when you reach negative values, since $\sigma(-x)$ is not defined. If you reach $d(0)$, then $\sigma(0) = n$ for the sake of this formula.

Euler found the sequence involved, 1, 2, 5, 7, 12, 15, 22, 26, 35, 40, 51, 57, 70, . . . by multiplying out,

$$(1 - x)(1 - x^2)(1 - x^3)(1 - x^4) \ldots$$
$$= 1 - x - x^2 + x^5 + x^7 - x^{12} - x^{15} \ldots$$

The sequence can also be derived from the *pentagonal numbers*. Their formula is $\frac{1}{2}n(3n - 1)$, and their values for positive and negative values of n go like this:

n	−5	−4	−3	−2	−1	0	1	2	3	4	5
$\frac{1}{2}n(3n-1)$	−40	−26	−15	−7	−2	0	1	5	12	22	35

The sequence can also be constructed by noting its first differences:

1	2	5	7	12	15	22	26	35	40	51	57	70
	1	3	2	5	3	7	4	9	5	11	6	13

in which the counting numbers and the odd numbers from 3 alternate.

Here's an example of how it works:

$$\sigma(10) = \sigma(10 - 1) + \sigma(10 - 2) - \sigma(10 - 5) - \sigma(10 - 7)$$
$$= \sigma(9) + \sigma(8) - \sigma(5) - \sigma(3)$$
$$= 13 + 15 - 6 - 4$$
$$= 18$$

It is thought-provoking that although $\sigma(n)$ apparently depends on the factors of n, it can also be calculated via the factors of a selection of numbers less than n.

divisors and partitions

Even more surprisingly, almost exactly the same formula, with the same condition on stopping, gives the number of partitions of n:

$$p(n) = p(n - 1) + p(n - 2) - p(n - 5) - p(n - 7) + p(n - 12)$$
$$+ p(n - 15) - p(n - 22) - p(n - 26) + p(n - 35) + p(n - 40) \ldots$$

The only difference is that if you reach $p(0)$ this is given the value 1.

Here is another connection: take any number, and write down its partitions into an *odd* number of *different* positive integers, for example,

$$11 = 1 + 2 + 8 = 1 + 3 + 7 = 1 + 4 + 6 = 2 + 3 + 6 = 2 + 4 + 5 = 11$$

Add the first terms: $1 + 1 + 1 + 2 + 2 + 11 = 18$.

Now partition 11 into even numbers of different positive integers:

$$11 = 1 + 10 = 2 + 9 = 3 + 8 = 4 + 7 = 5 + 6 = 1 + 2 + 3 + 5$$

Add the first terms: $1 + 2 + 3 + 4 + 5 + 1 = 16$. The difference, $18 - 16 = 2$, which is $d(11)$. This is a general result. (Bing, Fokkink, and Fokkink 1995)

curiosities of $\sigma(n)$

- The equation $\sigma(n) = \sigma(n + 1)$ has only nine solutions in positive integers less than 10,000. They are $n = 14$, 206, 957, 1334, 1364, 1634, 2685, 2974, and 4364.

- The equation $\sigma(n) + 2 = \sigma(n + 2)$ is satisfied whenever n and $n + 2$ are a prime pair, but there are three other solutions for n under 9998: $n = 434$, 8575, and 8825. (Makowski 1960)
- The values of $\sigma(n)$ peak at those n that have many small factors. The peak at $\sigma(60) = 168$ beats the previous peak of $\sigma(48) = 124$ by 44. The next record-breaking difference is between the peaks at $\sigma(108) = 280$ and $\sigma(120) = 360$.

prime factors

- The probability that the greatest prime factor of a random integer n is greater than $\sqrt{n}$ is log 2.
- The probability that a number N, chosen at random, has a prime factor between N^a and $N^{a(1+e)}$ is approximately equal to e, independently of the size of a, provided that e is small. (Riesel 1994, 161)
- The group 64, 65, 66 is the smallest triple of integers with, respectively, one, two, and three distinct prime factors. The next is 103, 104, 105, followed by 163, 164, 165 and 193, 194, 195.
- If n is greater than 239, then the largest prime factor of $n^2 + 1$ is at least 17. (Caldwell, *Prime Pages*)
- The number 140 is the start of the smallest sequence of seven consecutive integers each with an even number of prime factors.
- The number 170 is the start of the same record for an odd number of prime factors. (Honaker: Caldwell, *Prime Pages*)
- Six consecutive integers, beginning with 788, are divisible by the first six prime numbers, respectively.
- The triples 3, 4, 5 and 8, 9, 10 are the smallest non-overlapping sets of consecutive numbers sharing the same set of prime factors: 2, 3, and 5. Erdös has conjectured that there is only a finite number of such sets.

divisor curiosities

The pair $230 = 2 \cdot 5 \cdot 23$ and $231 = 3 \cdot 7 \cdot 11$ are the first pair of consecutive integers with three prime distinct factors each. The next pair is $285 = 3 \cdot 5 \cdot 19$ and $286 = 2 \cdot 11 \cdot 13$. The first triple of such consecutive integers is $1309 = 7 \cdot 11 \cdot 17$, $1310 = 2 \cdot 5 \cdot 131$, and $1311 = 3 \cdot 19 \cdot 23$.

See Appendix B [arithmetic functions]; factors of given form; factors, prime; probability; [Ramanujan's] highly composite numbers; squarefree numbers

economical numbers

If a number written as the product of its prime factors has no more digits than the number itself, the number is labeled economical. The sequence of economical numbers starts: 2, 3, 5, 7, 10, 11, 13, 14, 15, 16, 17, . . .

If it requires fewer digits, it is frugal; if it requires the same number, equidigital; and if it requires more, it is extravagant.

Frugal numbers, not surprisingly, are relatively scarce. According to Santos and Pinch the numbers of each kind for $2 \leq n \leq 500,000,000$ are:

frugal	1,455,952
equidigital	86,441,875
extravagant	412,102,173

Five sequences of seven economical numbers start at 157, 108749, 109997, 121981, and 143421. There is a sequence of length nine starting at 1034429177995381247. Pinch has checked up to 10^6 and only found *pairs* of consecutive frugal numbers, including $4374 = 2 \cdot 3^7$ and $4375 = 7 \cdot 5^4$.

However, Pinch has also proved that if **Dickson's conjecture** is correct, then there are sequences of consecutive frugal numbers of any specified length. (Caldwell, *Prime Pages*)

Electronic Frontier Foundation

The Electronic Frontier Foundation, cofounded by John Gilmore, offers large cash prizes in its "EFF Cooperative Computing Awards," which are offered to "encourage ordinary Internet users to collaborate to solve huge scientific problems." The emphasis is on cooperation.

They have already awarded $50,000 for the first prime number of more than one million digits, awarded on April 6, 2000, for the discovery by Nayan Hajratwala of the 38th **Mersenne prime**, $2^{6972593} - 1$, which actually has more than two million digits, as part of the **GIMPS** project.

Larger prizes are offered for larger record-breaking primes:

10,000,000 digits	$100,000
100,000,000 digits	$150,000
1,000,000,000 digits	$250,000

The EFF board will award the prizes based on the recommendation of the computing award advisory panel, including Curt Noll and Chris Caldwell. There are rules for distributing the prize money between the members of a collaborative group, such as GIMPS.

If you want to take part, visit www.eff.org/awards/coop.html.
See GIMPS, Mersenne primes

elliptic curve primality proving

The elliptic curve method for factorization and primality testing depends on an unusual property of a special type of curve. Elliptic curves were first studied in attempts to find the arc-length of an ellipse. Their equations have the form,

$$y^2 = x^3 + ax + b$$

Provided $4a^3 + 27b^2$ is not zero, the curve looks like the figure on page 57; cutting the *x*-axis at three points, the real roots of $x^3 + ax + b = 0$.

The unusual property is that we can "add" any two points on the curve to get a third point. We do it by drawing a line through the points we wish to add, such as *A* and *B*, and finding where this line meets the curve again, at *C'*. We then reflect *C'* in the *x*-axis to obtain the "sum" $A + B = C$.

There seems to be an exception to this possibility when the points *A* and *B* lie on a vertical line. However, we get around that apparent difficulty by the typically mathematical tactic of defining a *point at infinity*, denoted by ∞.

This process of "addition" has all the properties of a *mathematical group*, so we can drop the quotation marks and simply think of *group addition* defined on the points of the curve, with these group properties:

1. If *A* and *B* are points on the curve, then $A + B = C$ is on the curve.

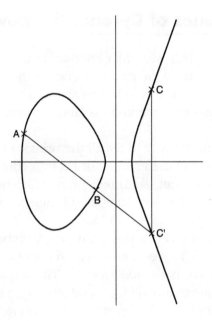

2. $(A + B) + C = A + (B + C)$. This is the *associative* property.

3. There is an *identity* element such that when added to any point A, the result is A. In this case the identity is the point at infinity, ∞, because according to our definition, $A + \infty = \infty + A = A$.

4. There is an inverse for each point, so that given A we can find A' such that $A + A' = \infty$. This is so because A' is simply the point where the vertical line through A cuts the curve again.

The properties of this group can now be used to test primality and to find factors. Readers who wish to go further into this remarkable method will find excellent accounts in Bressoud (1989) and Crandall and Pomerance (2001).

emirp

An emirp is a prime number that gives a different prime when the order of the digits is reversed. Palindromic primes are not included.

The sequence of emirps starts: 13, 17, 31, 37, 71, 73, 79, 97, 107, . . .

Eratosthenes of Cyrene, the sieve of

Eratosthenes (271–194 BC) was chief librarian at the famous Library of Alexandria. He calculated the circumference of the earth—252,000 stadia or about 24,662 miles, which was amazingly (and somewhat fortuitously) accurate—and invented his famous *sieve* for finding the prime numbers.

This is the most efficient way to list all the primes up to a few million: simply list them, and then strike out every second number; 3 is the first number missed out, so leave 3 but strike out every third number thereafter; 5 is the next number still missed out, so leave 5 and strike out every fifth number, and so on.

By arranging the numbers in this array, the numbers struck out form straight lines, which saves time spent on counting forward, and is a check that you have not made an error. The sequence of prime numbers appears, without any division and almost no multiplication. To sieve all the primes up to N, you stop when you reach the largest prime less than or equal to $\sqrt{N}$, which does need one multiplication. The process of striking out also reveals at least one of the factors of each composite number.

Basically the same method can be used to find all the primes in any interval.

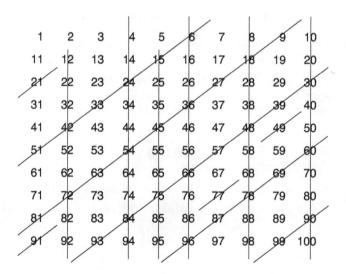

Schuler has calculated that 90.0037183% of all numbers greater than 257^2 are composite, with a smallest prime factor less than or equal to 257.

By cunning variations on Erastosthenes's basic idea, much more sophisticated sieve methods can be used. Viggo Brun introduced modern sieves in 1920, in a paper titled "The Sieve of Erastosthenes and the Theorem of Goldbach." He proved all these results:

1. If n is sufficiently large, then between n and $n + \sqrt{n}$ there is a number with at most eleven prime factors.
2. Every sufficiently large even number can be expressed as the sum of two numbers each having no more than nine prime factors.
3. There exists infinitely many pairs of numbers, having difference 2, of which the number of prime factors does not exceed nine.
4. For sufficiently large x, the number of prime twins not exceeding x does not exceed $100x/\log^2 x$.

These results are very far from being the best possible, but as so often in number theory, weak results are (mathematicians hope) a step in the right direction. The strongest version of (1.) is that there is a prime between n and $n + \sqrt{n}$. Chen proved in 1975 that there is a **semiprime** between n and $n + \sqrt{n}$. Point (2.) is a weak version of **Goldbach's conjecture**, and (3.) is a weak version of the **twin primes** conjecture. (Greaves 2001)

See lucky numbers

Erdös, Paul (1913–1996)

> God may not play dice with the universe, but something strange is going on with the prime numbers.
>
> —*Erdös (Mackenzie 1997)*

> Every human activity, good or bad, *except mathematics*, must come to an end.
>
> —*Erdös (Bollobás 1998)*

Paul Erdös was one of the greatest mathematicians of the twentieth century and one of the most prolific of all time, writing more than fifteen hundred papers, with nearly five hundred collaborators, whom he met as he wandered across the mathematical and geographical world, stopping with friends or favored universities where he would announce, "My brain is open!" He contributed especially to number theory, set theory, graph theory, and combinatorics, which fitted his cast of mind, and to probabilistic number theory, where the primes are treated as if they were **random**.

He claimed, "A mathematician is a machine for turning coffee into theorems," and he spent his adult life doing just that, working on problems up to within hours of his death. He didn't limit himself to coffee, however. He also, in later life, took amphetamines. Erdös often stayed with his close friend Ronald Graham and his wife, the mathematician Mei Fang, and Graham once bet him $500 that he could not give up amphetamines for a month. Graham lost. As Paul Hoffman tells the story, Erdös told him, "You've showed me that I'm not an addict. But I didn't get any work done. I'd get up in the morning and stare at a blank piece of paper. I'd have no ideas, just like an ordinary person. You've set mathematics back a month," and he went back to his pills. (Hoffman 1998, 16)

Erdös was also, perhaps, the most prolific problem poser ever. Not infrequently he offered prize money, from a few dollars up to $1,000 or more for the solution to problems that intrigued him, the value of the prize indicating his judgment of its difficulty. For example, he offered a prize of $10 for an answer to the question: can $\sigma(n)/n$ for a **weird number** be as large as we choose?

For harder problems he offered more money. In 1936, Erdös and Turán conjectured that any increasing sequence of integers that tends to infinity no faster than some arithmetic progression contains arbitrarily long arithmetic progressions. Szemerédi proved that this is true, in 1974, and won Erdös's prize of $1,000, the largest sum he ever had to actually pay out. (Bollobás 1998, 232)

Here is a related but harder problem. Suppose you have an infinite sequence of integers, $1 \leq a_1 < a_2 < a_3 < a_4 \ldots$ and suppose that the sum of their reciprocals, $1/a_1 + 1/a_2 + 1/a_3 + 1/a_4 + \ldots$ does *not* converge. This means, very roughly speaking, that there are "quite a lot of" the numbers. Must the sequence contain arbitrarily long arithmetic progressions?

Since the sum of the reciprocals of the prime numbers converges, an answer "yes" to this conjecture would prove, as a by-product, that there are arithmetic progressions of primes as long as we choose. No wonder that Erdös offered a prize of $3,000 in 1983, or that the conjecture has still not been settled, or that he remarked of the harder demand that the primes should be consecutive, "This conjecture is undoubtedly true but is completely unattackable by the methods at our disposal." (Erdös and Dudley 1983)

In 1990, in *A Tribute to Paul Erdös*, he presented "Some of my favourite unsolved problems," including this one: if $d_n = p_{n+1} - p_n$ is the difference between successive primes, are there an infinity of n for which $d_n < d_{n+1} < d_{n+2}$? Erdös was so certain this conjecture, which he could not prove, was true that he offered $100 for a proof but $25,000 for a disproof! (Baker, Bollobás, and Hajnal 1990, 469)

One day, Helmut Maier was giving Erdös a lift into town and mentioned a theorem he had just proved. " 'Maybe I offered a prize for that,' said Erdös, so of course they had to go to the library to check. Yes, indeed, Erdös had offered $100 for a proof, and he paid up at once. 'That's a pretty expensive taxi ride,' remarked Maier, and Erdös roared with laughter." (Pomerance, quoted in Seife 2002)

Erdös was also a supporter of elementary methods: this does *not* mean *simple* methods, but methods that do not use calculus. His very first paper, published when he was nineteen, used elementary methods that an undergraduate could understand to prove **Bertrand's postulate**.

In 1949 he and Atle Selberg found an elementary proof of the **prime number theorem** that did without the analytical tools used by Hadamard and Vallé Poussin. G. H. Hardy had previously said that such a proof "seems to me extraordinarily unlikely." The prime number theorem, Hardy thought, was extremely "deep," but an elementary proof would show that this was false, and in consequence it would be necessary for "the books to be cast aside and for the theory to be rewritten." Curiously, the proof by Erdös and Selberg did not have this effect at all. (Hardy 1966)

his collaborators and Erdös numbers

Because he collaborated with so many colleagues, the idea of Erdös numbers was invented. Mathematicians who have themselves collaborated with Erdös in writing a paper have Erdös number 1. Those who have collaborated with an Erdös number 1 colleague have Erdös number 2, and so on.

In 1996, 462 coauthors were listed, starting with George Szekeres in 1934 with five joint papers and Paul Turán, also 1934, with thirty.

By February 1999 the list had grown to 492 coauthors, and in February 2004 it was 509, partly because, like Euler, his papers have continued to be published after his death.

It is a well-known claim, sometimes called "Six Degrees of Kevin Bacon," that anyone can be connected to anyone else in the world by at most six links, so it is no surprise that many famous mathematicians and even physicists, some now dead, can be connected to Erdös and have their own Erdös number. Richard Dedekind (1831–1916) has Erdös number 7; Albert Einstein is a 2, Werner Heisenberg and Paul Dirac are both 4; the economist Kenneth Arrow is a 3, Noam Chomsky is 4, Stephen Hawking is also 4, and Bill Gates is a 4. Ahmad Chalabi, who "helped" Bush and Blair to "justify" the attack on Iraq, is a 6.

Here are just a handful, literally, of Erdös's many results and conjectures related to prime numbers:

There are infinitely many pairs of consecutive powerful numbers, but Erdös conjectured that there cannot be three consecutive **powerful numbers**.

Erdös's $n - 2^k$ conjecture: if you subtract from n all the powers of 2 less than n, when are the results all prime? Erdös conjectured that n must be 4, 7, 15, 21, 45, 75, or 105. Uchiyama and Yorinaga have verified this up to 2^{77}.

Erdös also conjectured that the same set of integers, $n - 2^k$, are **squarefree** for infinitely many n.

There are infinitely many primes, p, such that every even number less than $p - 2$ can be expressed as the difference of two primes less than or equal to p. For example, if $p = 23$,

$2 = 19 - 17$	$4 = 17 - 13$	$6 = 17 - 11$	$8 = 19 - 11$	$10 = 17 - 7$
$12 = 19 - 7$	$14 = 19 - 5$	$16 = 19 - 3$	$18 = 23 - 5$	$20 = 23 - 3$

This is a case where it would quite nice if 1 were counted as a prime: then $22 = 23 - 1$, obviously!

If p_n is the nth prime, as usual, this sequence converges:

$$-\tfrac{1}{2} + \tfrac{2}{3} - \tfrac{3}{5} + \tfrac{4}{7} - \tfrac{5}{11} + \tfrac{6}{13} - \tfrac{7}{17} + \tfrac{8}{19} - \tfrac{9}{23} \ldots$$

See conjectures; good primes; randomness

errors

In the eighteenth century, ideas of mathematical rigor were not well developed, and a mathematician as brilliant as **Euler** could use, for example, divergent series and draw conclusions that were usually correct, because his intuition was so powerful, but also sometimes wrong. So he claimed to have proved a version of the **prime number theorem**, but erred because a series was not absolutely convergent. (Erdös and Dudley 1983)

Making conjectures about the primes is especially easy and especially tempting, so in the history of the primes many **conjectures** have turned out to be false. There have also been errors of calculation.

Carmichael thought in 1906 that he had proved that the values of **Euler's totient function**, $\phi(n)$, never occur once only, but the "proof" contained an error, so in 1922 he published a correction and presented the problem as a conjecture—unproved to this day. Then in 1948 he published a correction of another claim in his original paper, and in 1949 he published a list of misprints in the 1948 paper! (Klee 1969)

Tables of primes prior to the advent of modern computers inevitably contained errors. Kulik constructed a table of the factors of numbers up to 100,330,200, excluding multiples of 2, 3, and 5, titled *Magnus Canon Divisorum pro omnibus numeris per 2, 3 et 5 non divisibilus, et numerorum primorum interfactentium ad millies centena millia accuratius as 100330201 usque.* He spent twenty years of life on this mammoth task, and when he died the manuscript was deposited in the Academy of Sciences in Vienna. Unfortunately, **D. N. Lehmer** found many mistakes in Kulik's work. To add insult to injury, the second of the eight volumes was already missing when D. N. Lehmer checked the manuscript, and has never been found.

The Danish mathematician Bertelsen claimed in 1893 that there are 50,847,478 primes less than 10^9. In 1959, **D. H. Lehmer** gave the correct figure, 50,847,534.

Euler's many volumes of collected papers contain a number of errors. He once announced that 1000009 was prime, only to realize his mistake later. We should sympathize: Euler was seventy at the time, and had been blind for years. (Caldwell, *Prime Pages*)

Stieltjes claimed in 1885 that he had proved the **Riemann hypothesis**, but he didn't publish his proof and died soon after in 1894 without doing anything further to support his claim. It is *very* likely that his proof contained an error—or did he see something that everyone since has missed? (Apolstol 2000, 10)

Carmichael and Mason listed fifty multiperfect numbers, but they mistakenly thought that $137561 = 151 \cdot 911$ and $485581 = 277 \cdot 1753$ were prime. (Guy 1994, 49)

More recently, a number of claimed results and proofs have turned out to be wrong. You might think that mathematicians would be just too careful to expose themselves to public humiliation, but, to be fair, current proofs are getting more and more complex and with the best will in the world, mistakes are easy to make.

Andrew Wiles announced in 1994 that he had proved **Fermat's Last Theorem**. The brilliant solution of a centuries-old problem made the front pages—only for a gap to be spotted, which he and his colleague Richard Taylor took more than a year to fill.

Carmichael's totient function conjecture mentioned above claims that the values of the **Euler totient (phi) function** are never unique. Filip Saidak claimed to have proved the conjecture in 1997, but his proof was erroneous.

Pogorzelski in 1977 thought he had proved **Goldbach's conjecture**, but "his proof is not generally accepted" by other number theorists. (Weisstein, *MathWorld*: "Goldbach Conjecture")

Dan Goldston and Cem Yildirim announced in 1985 that they had made a giant step toward solving the **twin primes** conjecture. Soon after, Andrew Granville and K. Soundararajan found an error. Goldston and Yildirim's conclusion depended on the size of the error term in one of their formulae. They had believed that the error term was bound to be especially small, making the formula especially accurate. It wasn't.

See Mersenne primes; Fermat; Polignac or obstinate numbers; Wiles

Euclid (c. 330–270 BC)

The Greek mathematician Euclid is most famous for writing his *Elements*, which systematically presented the elementary geometry of his day, starting with axioms and postulates and then proving a long sequence of theorems. It was so popular that it was still being used as a school textbook in the early twentieth century. However, the *Elements* also included basic number theory, in Book VII, propositions 22–32, on prime numbers, and propositions 33–39, on least common multiples, and in books VIII and IX.

unique factorization

The fundamental theorem of arithmetic states that each positive integer can be expressed as the product of primes in essentially one way only.

For example, $4773 = 3 \cdot 37 \cdot 43$
and, $6111 = 3^2 \cdot 7 \cdot 97$

This conclusion is not as obvious as it seems. Suppose that we consider only the even integers,

$$2, 4, 6, 8, 10, \ldots$$

We can factorize 8 into $2 \cdot 4$, and similarly, $20 = 2 \cdot 10$. However, 10 has no factors within the set, because 1 and 5 are not in the set. The "primes" within the set of even numbers are 2, 6, 10, 14, . . . (double the usual primes) and the composites are 4, 8, 12, . . .

It now turns out that $60 = 2 \cdot 30$ and $60 = 6 \cdot 10$, two distinct factorizations into "prime" numbers!

Another example consists of the ordinary integers with multiples of $\sqrt{6}$ added. So some sample integers in the set are $1 + \sqrt{6}$, $7 - 2\sqrt{6}$, $4 + 3\sqrt{6}$, and so on.

In this system of numbers it can be proved that 2 and 5 are prime, and so $10 = 2 \cdot 5$ is a prime factorization of 10. But we might also notice that $10 = (4 + \sqrt{6})(4 - \sqrt{6})$, which is another factorization, in which neither 2 nor 5 divide $4 + \sqrt{6}$ or $4 - \sqrt{6}$.

The fundamental theorem was eventually stated, clearly and explicitly and rather late in the day, by **Gauss**, but Euclid got very close. Book IX, proposition 14, states, "If a number be the least that is measured by prime numbers, it will not be measured by any other prime number except those originally measuring it." A number was "measured" by another if the other divided it without remainder, so this sounds like the fundamental theorem of arithmetic. Proposition 30 of Book VII claims that if a prime measures the product of two integers, then it must measure at least one of the integers. (Collison 1980)

$\sqrt{2}$ is irrational

No matter how many prime factors an integer n has, n^2 will have an *even* number. So the ratio of any two squares of integers also has an even number of prime factors, and since factorization into primes is unique, there is no way that this can be changed to an odd number of prime factors.

So *no* integer with an *odd* number of prime factors can be the square of a rational number. In particular, $\sqrt{2}$, $\sqrt{3}$, $\sqrt{5}$, $\sqrt{7}$, $\sqrt{11}$, $\sqrt{13}$, and $\sqrt{17}$ are all *irrational*, as are all the square roots of the following primes. (And so are the cube roots, 4th roots, 5th roots, etc.)

Strangely, the Greek Theodorus, Plato's tutor, followed Euclid and proved that the roots from $\sqrt{3}$ up to $\sqrt{17}$ were irrational—and then stopped!

Euclid and the infinity of primes

Euclid in Book VII of his *Elements* defined a prime number as "that which is measured by unity alone." In Book IX he proved that the number of primes is infinite, like this: start with the primes from 2 to 11, and calculate,

$$2 \cdot 3 \cdot 5 \cdot 7 \cdot 11 + 1 = 2311$$

The answer cannot be divisible by 2, 3, 5, 7, or 11, and so it is either prime itself or has at least two prime factors. (In fact it is prime.) Either way, we have produced at least one prime greater than 11.

Now assume that there are only a finite number of primes. Do the same, multiplying them all together and adding 1. Once again we shall create at least one new prime—which contradicts the assumption that the number of primes is finite. Therefore it is infinite.

The same idea can be used to start from any set of primes and find a prime not in the set. For example, we can calculate that,

$$2 \cdot 5 + 11 \cdot 17 = 217$$

Then either 217 is prime, or it has a prime factor that is not 2, 5, 11, or 17. In fact $217 = 7 \cdot 31$.

Or we could take the first four primes, 2, 3, 5, and 7, and form these sums and differences:

$$
\begin{array}{ll}
2 \cdot 3 \cdot 5 \pm 7 = 23, 37 & 3 \cdot 5 \cdot 7 \pm 2 = 103, 107 \\
5 \cdot 7 \cdot 2 \pm 3 = 67, 73 & 7 \cdot 2 \cdot 3 \pm 5 = 37, 47 \\
5 \cdot 7 \pm 2 \cdot 3 = 29, 41 & 3 \cdot 7 \pm 2 \cdot 5 = 11, 31 \\
3 \cdot 5 \pm 2 \cdot 7 = 1, 29 &
\end{array}
$$

Since 8, 9, and 10 are obviously divisible by either 2 or 3, any answer that is less than 11^2 must be prime, or unity.

The same calculation using 2^2, 3, 5, and 7 produces unity, plus the primes 13, 23, 41, 43, 47, 53, 67, 79, 89, 101, 109, 137, and the product of two new primes, $143 = 11 \cdot 13$.

consecutive composite numbers

Euclid's idea also suggests how we can find strings of consecutive composite numbers as long as we like. This expression,

$$1 \cdot 2 \cdot 3 \cdot 4 \cdot 5 \cdot 6 \cdot \ldots \cdot N + x$$

will be composite if x is any number from 2 to N. So if we want 100 consecutive composite numbers, $101! + x$, $x = 2$ to 101 will do.

primes of the form 4n + 3

We can use Euclid's method slightly adapted to prove that there is an infinity of primes of the form $4n + 3$.

The product of several numbers of the form $4n + 1$ is always of the same form, and the product of an *odd* number of numbers of the form $4n + 3$ is also of the form $4n + 3$: but the product of an *even* number of numbers of the form $4n + 3$ is of the form $4n + 1$.

Therefore, if we multiply, say, the first six terms in the $4n + 3$ sequence,

$$3 \quad 7 \quad 11 \quad 19 \quad 23 \quad 31 \quad 43 \quad 47 \quad \ldots$$

and add *2*: $3 \cdot 7 \cdot 11 \cdot 19 \cdot 23 \cdot 31 + 2$

we get a number of the form $4n + 3$ that is not divisible by 3, 7, 11, 19, 23, or 31. Its prime factors cannot all be of the form $4n + 1$, so at least one factor must be a "new" prime of the form $4n + 3$. Call it p, and form the expression,

$$3^2 \cdot 7 \cdot 11 \cdot 19 \cdot 23 \cdot 31 \cdot p + 2$$

Once again we have a product of an even number of numbers, $4n + 3$, with a factor of that form. So there is an infinity of $4n + 3$ primes.

Matching methods show that there is an infinity of primes of the form $5n + 4$, $8n + 3$, $8n + 5$, and $8n + 7$. (Sierpinski)

a recursive sequence

Starting with any set of primes, we can multiply them together and add 1; we then find the smallest prime factor of the result and add it to the original list. Do this starting with 2 and we get this sequence:

$$2 \quad 3 \quad 7 \quad 43 \quad 13 \quad 53 \quad 5 \quad \ldots$$

Which primes eventually appear in this sequence? The first prime that has not yet appeared in any calculation of the sequence is 31.

Another way of showing that there is an infinity of primes is to produce an infinite sequence of relatively prime numbers. One example is the **Fermat numbers**, $F_n = 2^{2^n} + 1$. One of their properties is that $F_m - 2 = F_0 F_1 F_2 F_3 \ldots F_{m-1}$. Since all F_m are odd, F_m has no factor in common with any of its predecessors, and so has a "new" prime factor.

Euclid and the first perfect number

Euclid showed in the *Elements*, Book IX proposition 36, that if $2^n - 1$ is prime then $2^{n-1}(2^n - 1)$ is a **perfect number**, equal to the sum of all its proper divisors.

Euclidean algorithm

This is a method of finding the *greatest common factor* or *divisor* (GCD) of two numbers. If the algorithm ends in 1, the original numbers are coprime. Euclid described it in his *Elements*, Book VII.

Let's start with the numbers 4334 and 2838. By division, which in this case is just subtraction,

$$4334 = 2838 + 1496$$

Continuing with 2838 and 1496,

$$2838 = 1496 + 1342$$

And so on:
$$1496 = 1342 + 154$$
$$1342 = 8 \times 154 + 110$$
$$154 = 110 + 44$$
$$110 = 2 \times 44 + 22$$
$$44 = 2 \times 22 + 0$$

The final remainder is 0, so the previous remainder, 22, is the greatest common factor. In fact, $4334 = 2 \times 11 \times 197$ and $2 \times 3 \times 11 \times 43$, from which we see that the common prime factors are 2 and 11.

The algorithm works because the GCD divides all the remainders from 1496 onward. If the smallest remainder is 1, then the numbers are coprime. For example:

$$5077 = 6 \times 813 + 199$$
$$813 = 4 \times 199 + 17$$

$$199 = 11 \times 17 + 12$$
$$17 = 12 + 5$$
$$12 = 2 \times 5 + 2$$
$$5 = 2 \times 2 + 1$$

Since we end up with the unit, 1, they are relatively prime. We could have stopped as soon as we noticed that 17 and 12 are relatively prime.

Lamé proved in 1844 that the Euclidean algorithm takes longest when the inputs are consecutive **Fibonacci numbers**.

Euler, Leonhard (1707–1783)

Till now mathematicians have tried in vain to this day to discover some order in the sequence of prime numbers, and we have reason to believe that it is a mystery into which the human mind will never penetrate. To convince oneself, one has only to glance at the tables of primes which some people took the trouble to compute beyond a hundred thousand, and one perceives that there is no order and no rule.

—Euler

Leonhard Euler was one of the greatest mathematicians of all time, up there with Archimedes, Newton, and Gauss. Like them, he excelled in pure and applied mathematics and created concepts and methods of enduring significance.

He studied at the University of Basel where the brilliant Johann Bernoulli was a professor. Euler was allowed to visit him on Saturday afternoons to quiz him on anything he found puzzling.

Euler published his first mathematical work at age eighteen, the precocious start of an astonishing output that eventually filled dozens of volumes, and included acoustics; algebra; artillery and ballistics; astronomy, including lunar motion and the calculation of orbits; calculus, including differential equations and the calculus of variations; cartography and geodesy; demography; electricity and magnetism; geometry, including differential geometry; hydraulics and hydrodynamics; insurance; mechanics; music; navigation and shipbuilding; number theory; optics; probability; and statistics.

He also contributed to what we call recreational mathematics. He wrote three memoirs on **magic squares**, and solved the well-known problem of the Bridges of Königsberg.

Due to a remarkable memory—even in old age he remembered the whole of Vergil's *Aeneid* by heart, page by page—he produced nearly half of his work after becoming blind after a cataract operation in 1771. His works were still being published by the St. Petersburg Academy nearly fifty years after his death.

Euler was a universal mathematician combining brilliance of calculation including use of observation and induction with dazzling formal manipulation and the creation of many novel concepts and methods that had a lasting impact: for example, he championed the potential use of divergent series, which do not converge to a finite sum, and he discovered the famous formula $e^{i\pi} = -1$, which links complex numbers and the trigonometrical functions.

The theory of numbers was a perfect arena for his genius. If Fermat built the initial foundations of the modern theory of numbers, Euler extended them and added several stories.

He proved **Fermat's Last Theorem** for the case $n = 3$, and found many new **amicable numbers**. He gave new proofs of **Fermat's Little Theorem** and generalized it by introducing **Euler's totient function**; he discovered the law of **quadratic reciprocity**, though he could not prove it. He studied the representation of primes in the form $ax^2 + by^2$, and found a method of **primality testing**.

Euler extended Fermat's discoveries on the forms of factors of certain numbers. He showed that if $n = a^{2^m} + b^{2^m}$, then all its prime factors are either 2, or of the form $2^{m+1}k + 1$. A special case is that every prime factor of the **Fermat number** $2^{2^5} + 1$ must be of the form $64k + 1$. Using this fact, he proved that 641 divides F_5, and so demolished **Fermat's conjecture** that all the Fermat numbers are prime.

Euler's convenient numbers

In his search for better ways to prove primality, Euler invented in 1778, when he was already seventy years old and nearly blind, the idea of a convenient number or *numerus idoneus*. These are the numbers such that IF n can be represented in the form $x^2 + dy^2$ in just one way, with x, y, and d coprime, the n must be prime.

Mersenne and Frenicle de Bessy had both realized that there was a connection between a number being composite and being representable as the sum of two squares in more than one way.

For example, Frenicle de Bessy in a letter dated August 2, 1641, challenged Fermat to factorize 221 by using the fact that $221 = 10^2 + 11^2$ and $5^2 + 14^2$.

Euler was the first to develop a specific method. He found a total of sixty-five convenient numbers, including all the integers 1 to 13, and expressed his belief that there would be an infinite number of them, and searched for them up to 10,000, though the largest he could find was 1848, from which he proved that

$$18518809 = 197^2 + 1848 \cdot 100^2$$

is prime, a remarkable feat for that time. Even more remarkably, we now know that Euler's list of convenient numbers was very likely complete: there is at most one more to be discovered, and if it exists it is greater than 10^{65}—a tribute to Euler's extraordinary powers of calculation.

Euler found his set of convenient numbers by using the theorem that m is a convenient number if and only if every integer $m + n^2 < 4m$ is either an odd prime or double an odd prime, or the square of an odd prime or a power of 2. (Actually, it is now known that this criterion is necessary but may not be sufficient.)

He gave these figures to show that 13 is a convenient number:

$13 + 1^2 = 14$	double an odd prime
$13 + 2^2 = 17$	odd prime
$13 + 3^2 = 22$	double an odd prime
$13 + 4^2 = 29$	odd prime
$13 + 5^2 = 38$	double an odd prime
$13 + 6^2 = 49$	square of an odd prime

He also claimed many properties for his convenient numbers, though he did not prove them all, including these true facts:

- The form $mx^2 + ny^2$ is convenient if and only if $x^2 + mny^2$ is.
- If an integer $4k - 1$ is convenient, then so is $4(4k - 1)$; and if $3k - 1$ is convenient, then so is $9(3k - 1)$.
- If n^2m is convenient, then so is m.
- The only square convenient numbers are 1, 4, 9, 16, and 25.

There is another connection between Euler's convenient numbers and a theme that appears repeatedly from Diophantus and Fermat onward: the idea that a prime is of the form $x^2 + y^2$ if and only if it is of the form $4n + 1$, and similarly for other forms.

It turns out that primes of the form $x^2 + Ny^2$ can be characterized by a condition such as $p = 4n + 1$ if and only if N is a convenient number. (Ore 1948, 61–63) (Frei 1985)

the Basel problem

Euler became immediately famous in 1735 when he solved the Basel problem, so named because Jacob Bernoulli, writing from Basel, had begged in one of his books for a solution to a problem that so many great mathematicians, including Jacob, Johann, and Daniel Bernoulli and Leibniz and de Moivre had tried and failed to solve. Euler showed that the sum of the series,

$$1/1^2 + 1/2^2 + 1/3^2 + 1/4^2 + 1/5^2 + \ldots$$

is equal to $\pi^2/6$. He also showed that,

$$1/1^4 + 1/2^4 + 1/3^4 + 1/4^4 + 1/5^4 + \ldots = \pi^4/90$$

and
$$1/1^6 + 1/2^6 + 1/3^6 + 1/4^6 + 1/5^6 + \ldots = \pi^6/945$$

He later calculated that,

$$1 + 1/3^2 + 1/5^2 + 1/7^2 + 1/9^2 + \ldots = \pi^2/8$$

and $\quad 1/1^{26} + 1/2^{26} + 1/3^{26} + 1/4^{26} + 1/5^{26} + \ldots = \dfrac{2^{24}}{27! \cdot 76977927 \cdot \pi^{26}}$

Euler also had the brilliant idea of studying the general series,

$$1 + 1/2^n + 1/3^n + 1/4^n + 1/5^n + 1/7^n + 1/8^n + 1/9^n + \ldots$$

He noticed that this is the product of all these infinite series, one for each prime:

$$(1 + 1/2^n + 1/2^{2n} + 1/2^{3n} \ldots)(1 + 1/3^n + 1/3^{2n} + 1/3^{3n} \ldots)(1 + 1/5^n + 1/5^{2n} + 1/5^{3n} \ldots)(1 + 1/7^n + 1/7^{2n} + 1/7^{3n} \ldots) \ldots$$

Summing each of these geometric series, the sum original sum equals

$$\dfrac{1}{1 - \dfrac{1}{2^n}} \; \dfrac{1}{1 - \dfrac{1}{3^n}} \; \dfrac{1}{1 - \dfrac{1}{5^n}} \; \dfrac{1}{1 - \dfrac{1}{7^n}} \cdots$$

an infinite product of terms, $1/(1 - 1/p^n)$, where p ranges over all the prime numbers from 2 onward.

This is an extraordinary connection, between a *sum* involving all the integers, to a *product* involving only the prime numbers! What is more, both the sum and the product are *functions of n* that can be manipulated like any other functions—so here, in this remarkable transformation, Euler was creating a bridge between the prime numbers as discrete integers, and analysis that deals with continuous quantities, and so taking the theory of prime numbers into entirely new territory.

Today, mathematicians consider the *zeta* series:

$$\zeta(s) = 1/1^s + 1/2^s + 1/3^s + 1/4^s + 1/5^s + \ldots$$

The famous **Riemann hypothesis** relates to this series, in which the integer n has become the complex number s.

Euler's constant

The series $\frac{1}{1} + \frac{1}{2} + \frac{1}{3} + \ldots + \frac{1}{n}$ diverges as n tends to infinity. Euler showed not only that it is approximately log n, but also that

$$\frac{1}{1} + \frac{1}{2} + \frac{1}{3} + \ldots + \frac{1}{n} - \log n$$

tends to a constant, Euler's constant, denoted by the Greek letter gamma, γ, which he calculated to sixteen decimal places in 1781. It is approximately 0.5772156649015328860 . . . It is not known whether it is rational. If it is, and $\gamma = \frac{a}{b}$, then $b > 10^{244662}$.

This is one of the most mysterious constants in mathematics, which turns up in some unexpected places. For example: if a large integer n is divided by each integer k, $1 \leq k \leq n$, then the average fraction by which the quotients fall short of the next integer is not $\frac{1}{2}$, but γ.

It has been conjectured that if $M(n)$ is the number of primes $p \leq n$, such that the Mersenne number $2^p - 1$ is prime, then $M(n)/\log n$ tends to the constant $e^\gamma \cdot \log 2 = 2.56954 \ldots$ as n tends to infinity. (Finch 2003, 29)

There are many other series for γ. This is Dr. Vacca's:

$$\gamma = (\tfrac{1}{2} - \tfrac{1}{3}) + 2(\tfrac{1}{4} - \tfrac{1}{5} + \tfrac{1}{6} - \tfrac{1}{7}) + 3(\tfrac{1}{8} - \tfrac{1}{9} + \tfrac{1}{10} - \tfrac{1}{11} \ldots - \tfrac{1}{15}) + \ldots$$

Euler and the reciprocals of the primes

Euler proved in 1737 that the sum of the reciprocals of the primes

$$\tfrac{1}{2} + \tfrac{1}{3} + \tfrac{1}{5} + \tfrac{1}{7} + \tfrac{1}{11} + \tfrac{1}{13} + \tfrac{1}{17} + \ldots$$

diverges. He also claimed, in modern notation, that it diverged like the function log log n. This is so slow that more than 360,000 terms are needed for the sum to exceed 3.

The alternating sum of the prime reciprocals converges, since its value always lies between ½ and 0:

$$½ - ⅓ + ⅕ - ⅐ + 1/11 - 1/13 + \ldots = 0.2696063519\ldots$$

The sum of the squared reciprocals of the primes also converges, to 0.4522474200 . . . (Finch 2003, 95)

Euler's totient (phi) function

This is is the number of integers less than n and prime to it, denoted by $\phi(n)$. (By convention, $\phi(1) = 1$.) It is therefore also known as the phi function. Its first few values are:

N	1	2	3	4	5	6	7	8	9	10	11	12
$\phi(n)$	1	1	2	2	4	2	6	4	6	4	10	4

If p is prime, $\phi(p) = p - 1$, and $\phi(p^a) = p^{a-1}(p - 1)$.

Like the two divisor functions, $d(n)$ and $\sigma(n)$, $\phi(n)$ is multiplicative, meaning that if m and n are coprime, then $\phi(mn) = \phi(m)\phi(n)$, so if we can write n as the product of its prime factors, we can at once calculate $\phi(n)$. In particular, if n is the product of distinct primes p, q, r, then $\phi(n) = (p - 1)(q - 1)(r - 1), \ldots$, for example,

$$\phi(105) = \phi(3 \cdot 5 \cdot 7) = 2 \cdot 4 \cdot 6 = 48$$

There is a related divisibility property: if a divides b, then $\phi(a)$ divides $\phi(b)$.

Apart from $\phi(1)$ and $\phi(2)$, $\phi(n)$ is always even. If n has r distinct odd prime factors, then 2^r divides $\phi(n)$.

Not all even numbers, however, are values of $\phi(n)$. For example, $2 \cdot 7^k$ is never a value. (Schinzel: Ribenboim 1995) On the other hand, every factorial $k!$ is a value of $\phi(n)$. (Gupta: Ribenboim 1995)

Other numbers occur frequently as values of $\phi(n)$. The number 12 occurs six times, for $n = 13, 21, 26, 28, 36,$ and 42. The number 16 also occurs six times, and 24 occurs ten times.

The sequence of even non-values of $\phi(n)$ starts:

14, 26, 34, 38, 50, 62, 68, 74, 76, 86, 90, 94, 98, 114, 118, . . .
(Sloane A005277)

The sum of $\phi(d)$ over every d that divides N, is N. For example, the divisors of 30 are 1, 2, 3, 5, 6, 10, 15, and 30, and,

$$\phi(1) + \phi(2) + \phi(3) + \phi(5) + \phi(6) + \phi(10) + \phi(15) + \phi(30)$$
$$= 1 + 1 + 2 + 4 + 2 + 4 + 8 + 8 = 30$$

If a number n has many small prime factors, it will have very many divisors and $\sigma(n)$ will be relatively large and $\phi(n)$ will be relatively small. So it is not surprising that $\sigma(n)/n$ and $n/\phi(n)$ have the same order of magnitude, $6/\pi^2 < \sigma(n)\phi(n)/n^2 < 1$ if $n > 1$. (Annapurna 1938)

The sum $\phi(1) + \phi(2) + \phi(3) + \phi(4) + \ldots + \phi(n)$ is very roughly $3n^2/\pi^2$ when n is large.

D. H. Lehmer conjectured that if n is composite, then $\phi(n)$ never divides $n - 1$. No counter-example is known. If it exists it is greater than 10^{20} and has more than thirteen prime factors. (Ribenboim 1995, 37) However, $\phi(n)$ divides n if and only if the only prime factors of n are 2 or 3, or both.

If p is prime, then $\phi(p) = p - 1$ and $\sigma(p) = p + 1$ and so $\phi(p) + \sigma(p) = 2p$. This is also a sufficient condition: no composite number has this property. What about the equation $\phi(n) + \sigma(n) = 3p$? This has solutions, 312, 560, 588, 1400, 85632, . . .

The equation $\phi(n) + \sigma(n) = 4p$ has solutions 23760, 59400, 153720, and 4563000.

Richard Guy also notes that if n is prime, then $\phi(n)\sigma(n)$ is 1 less than a perfect square, and this also occurs for $n = 6, 22, 33, 44, 69, 76, \ldots$ (Guy 1997)

Carmichael's totient function conjecture

We have already noted among famous **errors** that **Carmichael** conjectured that, if $\phi(n) = k$, then there is another number, m, such that $\phi(m) = k$ also, and believed that he had proved this. His proof was faulty, but the conjecture is now known to be true for all values of $\phi(n)$ up to $10^{1360000}$.

Erdös did prove that if there is an integer m, for which $\phi(n) = m$ has k solutions, then there are infinitely many integers with the same property. (Erdös and Dudley 1983)

Harold Donnelly proved that if $\phi(n) = k$ does have a unique solution, then $2^2 3^2 7^2 43^2$ divides n. (Donnelly 1973)

It isn't known whether there are an infinite number of pairs, such as 15 and 16, such that $\phi(n) = \phi(n + 1)$. However, Sierpinski proved that there is at least one solution to the equation,

$$\phi(n) = \phi(n + k)$$

for every value of k. The first few solutions are:

$k =$	1	2	3	4	5	6	7	8	9	10
$n =$	1	4	3	8	5	24	5	16	9	20

It was proved by D. J. Newman that there must be a value of n for which $\phi(30n + 1) < \phi(30n)$, but actual tests showed that there is no such $n \leq 20,000,000$. Then Greg Martin found the smallest n that satisfies the inequality. It is

232 909 810 175 496 793 814 049 684 205 233 780 004 859 885 966
051 235 363 345 322 075 888 344 528 723 154 527 984 260 176 895
854 182 634 802 907 109 271 610 432 287 652 976 907 467 574 362
400 134 090 318 355 962 121 476 785 712 891 544 538 210 966 704
036 990 885 292 446 155 135 679 717 565 808 063 766 383 846 220
120 606 143 826 509 433 540 250 085 111 624 970 464 541 380 934
486 375 688 208 918 750 640 674 629 942 465 499 369 036 578 640
331 759 035 979 369 302 685 371 156 272 245 466 396 227 865 621
951 101 808 240 692 259 960 203 091 330 589 296 656 888 011 791
011 416 062 631 565 320 593 772 287 118 913 728 608 997 901 791
216 356 108 665 476 306 080 740 121 528 236 888 680 120 152 479
138 327 451 088 404 280 929 048 314 912 122 784 879 758 304 016
832 436 751 532 255 185 640 249 324 065 492 491 511 072 521 585
980 547 438 748 689 307 159 363 481 233 965 802 331 725 033 663
862 618 957 168 974 043 547 448 879 663 217 971 081 445 619 618
789 985 472 074 303 100 303 636 078 827 273 695 551 162 089 725
435 110 246 701 964 021 045 849 081 811 604 427 331 227 553 783
590 821 510 091 607 567 178 842 569 576 699 548 038 217 673 171
895 383 249 326 800 667 432 993 531 186 437 659 910 632 865 419
892 370 957 722 154 266 351 039 808 548 150 828 868 968 820 675
198 820 381 135 523 646 361 202 383 915 218 571 017 801 463 011
491 108 784 343 253 284 393 511 650 254 506 597 923 969 653 616
813 897 710 621 756 693 827 471 154 701 151 222 320 443 347 408
180 047 964 860

Once again, the conjecture based on empirical evidence that a solution to a problem does not exist is overturned by a solution that takes a very long time to appear! (Martin 1999)

curiosities of $\phi(n)$

- 30 is the largest integer with the property that every smaller integer that is prime to it is itself a prime.
- $n - \phi(n)$ is never equal to 10. The sequence of values never taken by $n - \phi(n)$ continues: (10) 26 34 50 52 58 86 100 . . .
- $\phi(n) \geq \sqrt{n}$ unless $n = 2$ or $n = 6$.
- Carl Pomerance noticed that $210 - p$ is prime for all primes between 105 and 210. It was proved in 1993 that this is the only number with this property.

See errors; Fermat's Little Theorem; strong law of small numbers; Littlewood's theorem; Sierpinski's $\phi(n)$ conjecture

Euler's quadratic

> O'Toole never fully comprehended what exactly was meant by the expression "quadratic prime." However he did understand, and was fascinated by, the fact that the string 41, 43, 47, 53, 61, 71, 83, 97, . . . , where each successive number was computed by increasing the difference from the previous number by 2, resulted in exactly forty consecutive prime numbers. The sequence ended only when the forty-first number in the string turned out to be a non-prime, namely $41 \times 41 = 1681$.
>
> —*Arthur C. Clarke and Gentry Lee*

Euler discovered in 1772 that the formula $x^2 - x + 41$ is prime for $x = 1$ to 40, and for many values thereafter. Then **Legendre** noticed in 1798 that the quadratic $x^2 + x + 41$ is prime for $x = 0$ to 39, and this is, ironically, now known as Euler's polynomial. In fact it is composite for $x = 40$, 41, 44, 49, 56, 65, 76, 81, 82, 84, 87, 89, 91, and 96 (Sloane A007634) and prime for all other values up to 100 inclusive. Of the first 1,000 values, 581 are prime. (Caldwell, *Prime Pages*)

Legendre also noticed that $2x^2 + 29$ is prime for $x = 0, 1, \ldots, 28$. In general, the polynomial $2x^2 + p$, with $p = 3$, 5, 11, or 29, gives prime values for $x = 0, 1, \ldots, p - 1$.

Euler's formula was used by Charles Babbage to demonstrate the capabilities of his Difference Engine. As he explained in a letter to Sir Humphrey Davy in 1822, the machine could calculate but not print, so an assistant had to write down the numbers as they were produced:

> [The Difference Engine] proceeded to make a table from the formula $x^2 + x + 41$. In the earlier numbers my friend, in writing quickly, rather more than kept pace with the engine; but as soon as four figures were required, the machine was at least equal in speed to the writer. (Williams and Shallit 1994, 488)

The values $x = -40$ to -1 produce the same set of prime values. By substituting $x - 40$ for x in the original formula, we get the quadratic $x^2 - 79x + 1601$, which produces eighty primes in sequence from $x = 0$ to 79, but they are duplicated.

There is a deep reason why the formula $n^2 + n + 41$ gives a prime for all values of n between 0 and 39, and a simpler reason why we might expect it to give an unusually large number of primes for any value of n. Here is the simple reason.

First, $n^2 + n$ is even, so we are not surprised that 41 is odd. Similarly, $n^2 + n \equiv 0$ or 2 (mod 3), and we notice that $41 \equiv 2$ (mod 3), so $n^2 + n + 41 \equiv 1$ or 2 (mod 3). Likewise, $n^2 + n \equiv 0$, 1, or 2 (mod 5), and we notice that $41 \equiv 1$ (mod 5), so $n^2 + n + 41 \equiv 1$, 2, or 3 (mod 5). Continuing with 7, and combining the results, we find that $n^2 + n + 41$ is never divisible by 2, 3, 5, or 7. More generally, if $x \equiv$ 11, 17, 41, 101, 137, or 167 (mod 210), then $n^2 + n + x$ will never be divisible by either 2, 3, 5, or 7.

The number of primes of the form $n^2 + n + x$ for $0 < n < 1000$, and these values of x are:

x	11	17	41	101	137	167
prime values of $n^2 + n + x$	288	366	582	453	339	285

("Prime Producing Polynomials," www.glasgowg43.freeserve.co .uk/primpoly.htm)

the Lucky Numbers of Euler

It is a theorem that no polynomial, $f(n)$, with integer coefficients can be a prime for all n or for all sufficiently large n, unless it is a constant. However, some polynomials do produce exceptionally high proportions of primes.

Prime numbers, p, such as 41, for which the quadratic $x^2 + x + p$ produces many primes have been called by Le Lionnais the Lucky

Numbers of Euler. For instance, $x^2 - 2999x + 2248541$ produces eighty different primes from $x = 1460$ to 1539. (Beiler 1966, 220)

As a result of computer searches, other forms of quadratics are known that give more primes. Beeger in 1938 discovered that $x^2 + x + 27941$ produces more primes among its first million values, from $x = 0$ onwards, than does $x^2 + x + 41$, 286128 to 261080. Gilbert Fung discovered,

$$103n^2 - 3945n + 34381 \text{ for } n = 0 \text{ to } 45$$

Russell Ruby found: $36x^2 - 810x + 2753$, which is prime for $x = 0$, $1, \ldots, 44$.

According to one of the **Hardy-Littlewood conjectures**, the density of primes among the values of these quadratic expressions as n tends to infinity will always be $c\sqrt{n}/\log n$, where c will vary with the expression. Euler's formula has $c = 3.3197732$, whereas the formula $n^2 + n + 132874279528931$ has $c = 5.0870883$.

Ruby and Fung also found the second record polynomial, $47x^2 - 1701x + 10181$, which produces forty-three distinct prime values for $0 \leq x \leq 42$. (Fung and Williams 1990)

The polynomial $41x^2 + 33x - 43321$ gives prime values for ninety values of x, $0 \leq x \leq 99$, but at most twenty-six of these are consecutive.

If one of the Hardy-Littlewood conjectures is true, then the quadratic

$$x^2 + x + 1712,32986,61656,08771$$

has a record-breaking asymptotic density, meaning that as x tends to infinity, it has the largest known proportion of primes among its values.

Sierpinski has proved that for any integer N, you can find a number c such that $x^2 + c$ has at least N prime values. It has since been proved that this holds for any polynomial of degree greater than 1, with integral coefficients. (Abel and Siebert 1993)

See Euler's constant; Euler and the reciprocals of the primes; Euler's totient (phi) function; Riemann hypothesis

factorial

The factorial of an integer, n, written $n!$, is the product of all the integers up to and including n, so $8! = 1 \cdot 2 \cdot 3 \cdot 4 \cdot 5 \cdot 6 \cdot 7 \cdot 8 = 40{,}320$.

An approximate formula for large factorials discovered by James Stirling (1692–1770) is one of the most remarkable in elementary mathematics:

$$n! \sim \sqrt{2}\sqrt{\pi}\, n^{n+\frac{1}{2}} e^{-n}$$

It is quite impressive even for small values of n. For example, $20! = 2{,}432{,}902{,}008{,}176{,}640{,}000$, and using an ordinary pocket calculator, the product on the right equals $2.4226 \cdot 10^{18}$.

factors of factorials

What power of (for example) 11 divides $203!$? **Legendre** discovered the answer in 1808. Represent 203 as the sum of powers of 11:

$$203 = 1 \cdot 11^2 + 7 \cdot 11 + 5$$

so that $203_{11} = 175$. Then the answer is $(203 - (1 + 7 + 5))/(11 - 1) = 19$.

The highest power of 2 that divides $2^m!$ is $2^{2^m - 1}$. (Lariviere)

The product of any n consecutive positive integers is divisible by $n!$.
   See also Pascal's triangle; Wilson's theorem

factorial primes

Factorial primes are of the form $n! \pm 1$. Both forms have been tested to $n = 10{,}000$:

- $n! + 1$ is prime for $n = 1, 2, 3, 11, 27, 37, 41, 73, 77, 116, 154, 320, 340, 399, 427, 872, 1477, 6380, 26951, \ldots$
- $n! - 1$ is prime for $n = 3, 4, 6, 7, 12, 14, 30, 32, 33, 38, 94, 166, 324, 379, 469, 546, 974, 1963, 3507, 3610, 6917, 21480, 34790, \ldots$

These are the five largest known factorial primes, with the year of their discovery, and their discoverer. (Caldwell, *Prime Pages*)

Search of the Next Prime of the Form n! + 1 and n! – 1
(from http://powersum.dhis.org/)

The purpose of this page is to coordinate effort to search for the next primes of the form n! + 1 and n! – 1.

At the moment our purpose is to search the range between 30,000 → 100,000.

All numbers in the range have been trial factored by Phil Carmody up to 6,320,124,029. The next step is perform a probable primality test for every remaining number. The number 35,000! is 143,845 digits long and the number 100,000! is 456,574 digits.

Please join us in the search!
News Flash I!
On May 16, 2002, Leon Marchal found 34790! – 1 is prime!
This prime is 142891 digits long.
News Flash II!
On May 22, 2002, Ken Davis found 26951! + 1 is prime!
This prime is 107707 digits long.

34790! – 1	142,891 digits	2002	Marchal, Carmody, Kuosa
26951! + 1	107,707 digits	2002	Davis, Kuosa
21480! – 1	83,727 digits	2001	Davis, Kuosa
6917! – 1	23,560 digits	1998	Caldwell
6380! + 1	21,507 digits	1998	Caldwell

These factorial primes are also the subject of an Internet-based search; see the box on this page.

factorial sums

The sums in this sequence,

$$
\begin{aligned}
n = 3 \quad & 3! - 2! + 1! && = 5 \\
n = 4 \quad & 4! - 3! + 2! - 1! && = 19 \\
n = 5 \quad & 5! - 4! + 3! - 2! + 1! && = 101 \\
n = 6 \quad & 6! - 5! + 4! - 3! + 2! - 1! && = 619 \\
n = 7 \quad & 7! - 6! + 5! - 4! + 3! - 2! + 1! && = 4421
\end{aligned}
$$

are all prime. Also $n = 8$ and $n = 10$ lead to primes, but when $n = 9$, the sum is $326981 = 79 \cdot 4139$.

From $n = 11$ to $n = 28$ there are just two more primes, so this looks like another case of the **strong law of small numbers**.

Keller has found that $160! - 159! + 158! - \ldots - 3! + 2! - 1!$ is prime. (Keller: Caldwell, *Prime Pages*)

The sum $1! + 2! + 3! + 4! + \ldots + n!$ is only a square for $n = 1$ or 3. (Koshy 2002)

factorials, double, triple . . .

The double factorial of n is $n!! = n(n - 2)(n - 4) \ldots$ and the triple factorial is $n!!! = n(n - 3)(n - 6) \ldots$ and so on, stopping before the first term that is either negative or zero.

The function $n!! + 1$ is even when n is odd, and contains very few primes when n is even. The prime values of $n!! + 1$ up to $n = 5898$ are just $n = 1, 2,$ and 518. The fourth and fifth primes of this form are $33416!! + 1$ and $37310!! + 1$. (Harvey: Caldwell, *Prime Pages*) Curiously, $n!! - 1$, to the same limit, is prime for all the values $n = 2, 4, 6, 8, 16, 26, 64, 82, 90, 118, 194, 214, 728, 842, 888,$ and 2328. (Ribenboim 1995, 27)

The largest known prime of the form $n!! - 1$ is $9682!! - 1$. (de Water: Caldwell, *Prime Pages*)

The largest known triple factorial prime is $34706!!! - 1$. (Harvey: Caldwell, *Prime Pages*)

factorization, methods of

Pick up a calculator, key in two numbers, and multiply them together, say, $1077 \times 3463 = 3729651$. Now ask a friend to tell you what numbers you multiplied! Tricky! Very tricky!

It isn't difficult because you used a calculator—if you were taught to do long multiplication in school you could work the product out for yourself on a scrap of paper in a few minutes at most. But to work *backwards*! Ah! There's the rub!

It is one of the curiosities of arithmetic that it is so easy to multiply numbers together, yet to factorize even quite a small number is far harder and cannot be done by any really quick and simple method.

Multiplication with a hand calculator is so easy it's trivial: but ordinary calculators do not have a "factor" key, and finding factors on a calculator is long-winded and tedious, or even impossible.

A correspondent once asked **Fermat** whether 100,895,598,169 is prime, and Fermat replied immediately that it is composite, the product of 898,423 and 112,303. How did he do it? No one knows, and the numbers are so large—prior to the age of electronic calculators—that it has been speculated that Fermat had a secret method of factorization, since lost.

Oliver Sacks describes in his book *The Man Who Mistook His Wife for a Hat, and Other Clinical Tales* the twins John and Michael, who, given a number of up to twenty digits, could say extremely quickly whether it was prime, though they had difficulty with elementary arithmetic. (Sacks 1985)

Frank Nelson Cole (1861–1926)

One of the most extraordinary meetings in the history of mathematics was described by E. T. Bell in *Mathematics: Queen and Servant of Science*:

> At the October, 1903, meeting in New York of the American Mathematical Society, Cole had a paper on the programme with the modest title, "On the Factorization of Large Numbers." When the chairman called on him for his paper, Cole—who was always a man of few words—walked to the board and, saying nothing, proceeded to chalk up the arithmetic for raising 2 to its 67th power. Then he carefully subtracted 1. Without a word he moved over to a clear space on the boards and multiplied out, by longhand:

$$193,707,721 \times 761,838,257,287$$

> The two calculations agreed. . . . Cole took his seat without uttering a word. Nobody asked him a question.

Only later did Cole admit that he had been working on this problem for the previous twenty years.

The simplest method is the sieve of **Erastosthenes**, which can be implemented very efficiently on a computer to find prime factors of numbers up to several million. In modern factorization methods, the idea of a sieve has been revived. The best-known algorithm for factoring large numbers is the general number field sieve.

The second method is trial division, which is the natural approach with a hand calculator. You notice from the last digit and the sum of digits that the number is not divisible by 2, 5, or 3, and then you start dividing by 7, 11, 13, 17, . . . Since at least one of the prime factors of n must be at most $\sqrt{n}$, you need only make a limited number of trials—but that number could be very large. To test 2053, for example, you need to divide by all the primes up to $\sqrt{2053} = 45.3 \ldots$

For a large number, this will take a long time! It would be simpler if you could know in advance what *kind* of factors you were looking for, in order to eliminate the rest. This can sometimes be done.

factors of particular forms

Euler proved that if a prime p divides $x^2 + y^2$ but does not divide both x and y, then p is the form $u^2 + v^2$. Similarly, if p divides $x^2 + 2y^2$ or $x^2 + 3y^2$, then under the same condition it must be of the forms $u^2 + 2v^2$ or $u^2 + 3v^2$, respectively.

Fermat discovered that the prime factors of $2^p - 1$ are all of the form $2kp + 1$, where k is a positive integer. So to factorize the **Mersenne number** $2^{11} - 1 = 2047$ we need only try the factors $22k + 1$ that are less than $\sqrt{2047}$ or roughly 45.

The sequence $22k + 1$ starts, 23, 45, 67, 89, . . . As it happens, the very first works: $2047 \div 23 = 89$.

Later, Euler proved that any prime factor of a **Fermat number**, $F_n = 2^{2^n} + 1$, if $n > 2$, is of the form $k \cdot 2^{n+2} + 1$.

Since primes of the form $4n + 1$ can be written as the sum of two squares in one way only, and primes of the form $4n + 3$ not at all, you can prove that a number is composite by writing it as a sum of two squares in at least two ways. Euler proved that 1,000,009 is composite by showing that it could be written as:

$$1{,}000{,}009 = 1000^2 + 3^2 = 235^2 + 972^2$$

The same fact also allows the factors to be calculated.

Conversely, some forms are always composite. Every term of the sequence of numbers $78557 \cdot 2^n + 1$ is divisible by one of the primes 3, 5, 7, 13, 19, 37, or 73. The number 78557 is probably the smallest with this property.

Fermat's algorithm

Fermat also introduced in 1643 another method, the first workable idea for factorizing large numbers that have no special form. He suggested adding perfect squares to the number, n, to be factored and testing to see if the result was a perfect square. If it is, then he had numbers a and b such that,

$$n + a^2 = b^2 \qquad \text{or} \qquad n = b^2 - a^2 = (b - a)(b + a)$$

Factors of Given Form

The prime factors of $2^n + 1$ are shown in this table:

n	$2^n + 1$	n	$2^n + 1$
1	3	9	$3^3 \cdot 19$
2	5	10	$5^2 \cdot 41$
3	3^2	11	$3 \cdot 683$
4	17	12	$17 \cdot 241$
5	$3 \cdot 11$	13	$3 \cdot 2731$
6	$5 \cdot 13$	14	$5 \cdot 29 \cdot 113$
7	$3 \cdot 43$	15	$3^2 \cdot 11 \cdot 331$
8	257	16	65537

The prime factor 3 appears first at $n = 1$ (marked in bold) and then at $n = 3, 5, 7, 9, \ldots$ The prime factor 5 appears first at $n = 2$ (marked in bold) and then at $n = 6, 10, 14, 18, \ldots$ The prime factor 11 appears at $n = 5$ and then at intervals of 10, and the prime factor 17 appears first at $n = 4$ and then at intervals of 8, which is $\frac{1}{2}(17 - 1)$.

If p appears for the first time as a factor of $2^n + 1$, then $p \equiv 1 \pmod{n}$. If $p \mid 2^n + 1$, then $p \mid 2^{p-1}(2^n + 1) = 2^{n+p-1} + 2^{p-1} = 2^{n+p-1} + 1 + (2^{p-1} - 1)$ and since $p \mid 2^{p-1} - 1$ by **Fermat's Little Theorem**, $p \mid 2^{n+p-1} + 1$. Therefore p is a factor at intervals of *at most* $n - 1$. (McLean 2002, 466–67)

Moreover, this method will find every pair of factors in theory. We simply calculate half the sum and half the difference of the factors. For example, $117{,}983 = 127 \cdot 929$:

$$\tfrac{1}{2}(127 + 929) = 528 \quad \text{and} \quad \tfrac{1}{2}(929 - 127) = 401$$

and $528^2 - 401^2 = 117{,}983$.

This illustrates, however, the difficulty with this method. You have to add a lot of squares in sequence to $117{,}983$ before you add 401^2. This method is only really efficient when you are confident that the factors are pretty equal in size. However, the same basic idea is used in more modern and sophisticated methods such as the quadratic sieve and the continued fraction algorithm.

Legendre's method

Legendre (1752–1833) considered a related idea. Given a number, N, that he wanted to factor, he took a prime, p, and searched for solutions of the congruence,

$$x^2 \equiv \pm p \ (\text{mod } N)$$

Carl Pomerance Recalls

"When I give talks on factoring, I often repeat an incident that happened to me long ago in high school. I was involved in a math contest, and one of the problems was to factor the number 8051. A time limit of five minutes was given. It is not that we were not allowed to use pocket calculators; they did not exist in 1960, around when this event occurred! Well, I was fairly good at arithmetic, and I was sure I could trial divide up to the square root of 8051 (about 90) in the time allowed. But on any test, especially a contest, many students try to get into the mind of the person who made it up. Surely they would not give a problem where the only reasonable approach was to try possible divisors frantically until one was found. There must be a clever alternate route to the answer. So I spent a couple of minutes looking for the clever way, but grew worried that I was wasting too much time. I then belatedly started trial division, but I had wasted too much time, and I missed the problem.

"The trick is to write 8051 as $8100 - 49$ which is $90^2 - 7^2$, so we may use algebra, namely, factoring a difference of squares, to factor 8051. It is 83 times 97." (Pomerance 1996)

In other words, he checked whether $\pm p$ was a **quadratic residue** (mod N). If it is, then it is also a quadratic residue modulo any prime factor of N, and this information reduces the number of possible prime factors that need to be considered. By finding more and more primes that are quadratic residues of N, more and more possible factors are eliminated; if he could eliminate all possible factors up to $\sqrt{N}$, then he knew that N was prime.

congruences and factorization

Maurice Kraitchik suggested an ingenious variation on Fermat. Instead of solving $n = x^2 - y^2$ you only try to solve the congruence,

$$x^2 = y^2 \ (\text{mod } n)$$

This means that n divides $x^2 - y^2 = (x - y)(x + y)$, so that if n divides neither $x + y$ nor $x - y$ then n must share factors with both $x + y$ or $x - y$.

How do you find likely values of x and y, however? That's the difficulty. If there were a simple and quick method of finding x and y, then factorization would be quick and easy after all, but there isn't.

So an element of luck comes in and the modern powerful methods for factorization that are based on Fermat's original idea all use random numbers. The result—ironically—allows large numbers to be factorized more quickly, and these factorizations, despite the use of random numbers, are *definite*: they are quite different from the *probabilistic* algorithms that are used to tell you *with a high probability* that a number is prime (or composite).

how difficult is it to factor large numbers?

Modern methods of factorization depend on advanced mathematical techniques, such as the **elliptic curve** method, which can be used to factor as well as prove primality.

Such powerful methods are necessary, because it takes about twice as much work to factorize $1000n$ as it does to factor n, so factorizing a 100-digit number is far, far harder than factorizing even an 80-digit number.

In 1970 it was still very difficult to factorize 20-digit numbers that did not have any especially convenient form, such $2^n - 1$. By 1980, the Brillhart-Morrison *continued fraction algorithm* allowed 50-digit numbers to be factorized easily.

In 1984 the Association for Computing Machinery presented a plaque to the Institute of Electrical and Electronics Engineers on the

---

Maurice Kraitchik (1882–1957)

Maurice Kraitchik was a Russian-born Belgian mathematician who wrote on the theory of numbers and (like **Lucas**) on recreational mathematics, including knight's tours, rational triangles and the Euler brick, **magic squares**, and magic tricks, including that known today as Total Destiny. Between 1931 and 1939 he edited the journal, *Sphinx: Revue Mensuelle des Questions Récréatives,* and he published *The Mathematics of Games* and *Mathematical Recreations* (1943/1960).

Kraitchik conjectured that if $2n - 1$ is prime, then $(2n + 1)/3$ is also prime, but this fails when $n = 89$. However, Kraitchik's conjecture has been improved to form the **New Mersenne conjecture**.

---

occasion of the IEEE centennial. It was inscribed with the prime factorization of the number $2^{251} - 1$ that was completed that year with the recently developed *quadratic sieve algorithm.*

In 1990 Carl Pomerance's quadratic sieve algorithm had doubled the length of factorizable numbers up to a record 116 digits, and by 1994 the quadratic sieve had been used to factorize the famous 129-digit RSA challenge number that Martin Gardner in a 1976 *Scientific American* column had judged would be safe for 40 quadrillion years.

In spring 1996, Pollard's number field sieve became the new champion, factorizing a 130-digit RSA challenge number in about 15% of the time required by the quadratic sieve. (Pomerance 1996)

Richard Mollin remarks that there has been a history of seriously overestimating the difficulty of factorizing integers. The surprising success of the **AKS algorithm for primality testing** suggests that perhaps much more powerful methods of factorization are awaiting discovery. If they aren't, there is always the possibility of more powerful machinery in the form of quantum computers.

quantum computation

Testing for primality and factorization of large numbers would be almost incomparably faster if quantum computation becomes a reality. Instead of calculating on a string of numbers, basically n binary bits, a quantum computer could operate simultaneously on an n-dimensional "cube" of 2^n binary bits.

The results are theoretically amazing. For example, Shor in 1994 discovered a polynomial time algorithm for factorizing integers, for a quantum computer. If it could be realized, the RSA cryptosystem and all similar systems would become useless. According to Adleman a DNA computer could perform 10^{20} operations per second, or 100 million times faster than a current supercomputer. (Mollin 2001, 267, 269)

See elliptic curve primality proving; factors of particular forms; Fermat; GIMPS; Mersenne numbers; primality testing

Feit-Thompson conjecture

The conjecture that there are no primes p and q for which $(p^q - 1)/(p - 1)$ and $(q^p - 1)/(q - 1)$ have a common factor. However, the counterexample $p = 17$, $q = 3313$ with a common factor of 112,643 was subsequently found by Stephens (1971). There are no other such pairs with both values less than 400,000. (Wells 1986)

Fermat, Pierre de (1607–1665)

> I have found a very great number of exceedingly beautiful theorems.
>
> —*Fermat*

Fermat was a lawyer by profession but also something of a classical scholar, fluent in Greek and Latin as well as Italian and Spanish, and a poet in Latin.

Fermat and Blaise Pascal founded the modern theory of probability, in their correspondence during the summer of 1654. Fermat also wrote on optics, and Fermat's principle, that light always follows the path that takes the shortest time, is named after him.

He developed a method of solving equations of the form $x^2 - ay^2 = 1$, now incorrectly called a Pell equation.

He also performed many integrations and differentiations, though he never reduced his arguments to a method, and so does not share with Newton and Leibniz the honor of creating the calculus.

His works were not published during his lifetime because he had an inexplicable aversion to publication, though he readily sent his results to friends. Explaining one of his discoveries to Mersenne, he wrote, "I would send you a proof, if I did not fear its being too long."

When he created his own form of analytic geometry (by 1636), it wasn't published, while Descartes published his in 1637, and Fermat's work in the theory of number was appreciated by almost no one until **Euler** rediscovered it.

Fermat was the first modern number theorist. He read Vieta, who had introduced a new symbolic algebra, and he studied the *Arithmetica* of **Diophantus**. Some of his most important conclusions, including his statement of **Fermat's Last Theorem**, were only preserved because his eldest son Samuel published an edition of Diophantus to which he added his *Observations on Diophantus* based on his father's marginal annotations.

Ironically, although he claimed to have a general method that lay behind his many results, Fermat never revealed it, apart from his one idea of "infinite descent." He also claimed to be able to prove what Diophantus had conjectured, that all positive integers are the sum of at most four squares. Unlike Diophantus, he only sought for solutions in integers, not fractions, which led him to emphasize prime numbers and divisibility.

In the absence of learned journals in which to share their results, mathematicians of that era had a custom of challenging each other to solve certain problems. In 1640, Frenicle de Bessy challenged Fermat to find a **perfect number** of twenty digits "or the next one following it." In tackling this problem, which was really about **Mersenne primes**, of the form $2^p - 1$, Fermat made three discoveries: that if $2^n - 1$ is prime, then n is prime also; that if p is an odd prime, then $2p$ divides $2^p - 2$; and that if p is prime, the prime factors of $2^p - 1$ are all of the form $2kp + 1$, where k is a positive integer. (Williams 1998, 36)

These were extremely significant discoveries. The first typically shows that if a number involving a power has a certain property, the index has a property also.

The third shows that we can say something very restricting about the factors of certain numbers, which at once makes them much easier to factorize. Fermat used it to show that 223 divides $2^{37} - 1$.

The second conclusion is the basis for **Fermat's Little Theorem** (to distinguish it from **Fermat's Last Theorem**).

Fermat's work eventually had a profound influence. When Euler discovered Fermat's results, he was prompted to try to prove many of them, with brilliant success, and when Augustin-Louis Cauchy's father, writing in 1812, wanted to encourage him in his efforts to become accepted, he wrote, "Your last paper on polyhedra made a deep impression on the Académie [des Sciences]. If you prove one of Fermat's theorems, [on polygonal numbers] the way will be wide open for you. The moment is favourable for you. Do not let it slip by."

Cauchy didn't. In 1815 he presented a sensational paper in which he proved Fermat's claim in a letter to Mersenne of 1636, that "Every number is the sum of three cubes, of four squares, and so on, indefinitely."

Fermat's Little Theorem

If p is prime then,

$$a^p \equiv a \pmod{p}$$

We can prove this by linking it to the **binomial theorem**, which says that $(a + 1)^p - a^p$ is equal to

$$\left(a^p + pa^{p-1} + p\,\frac{(p-1)}{2}a^{p-2} + \ldots + pa + 1 \right) - a^p$$

Since p is prime, all the binomial coefficients are divisible by p, and this expression is congruent to 1 (mod p).

$$(a + 1)^p - a^p \equiv 1 \pmod{p}$$

and so, $\qquad (a + 1)^p - (a + 1) \equiv a^p - a \pmod{p}$

So the value of $(a^p - a)$ does not depend on a, provided p is prime. So it must be equal to the value of $1^p - 1 = 0$. That is, $a^p \equiv a \pmod{p}$, which is what we wanted to show.

If p and a are coprime, then we can divide by p and conclude that

$$a^{p-1} \equiv 1 \pmod{p}$$

This looks like a promising basis for a primality test—certainly much better than **Wilson's theorem**, because a^{p-1} is easier to calculate than a factorial. Unfortunately there is a snag: there are many numbers, p, that are *not* prime but that satisfy the equation. These are called **pseudo-primes** and it is their existence that makes primality testing so much harder. To test a number, q, for primality using Fermat's Little Theorem, we can calculate a^{q-1} for some value of a, divide by q, and see if the remainder is 1. If it is not, then q is composite. If the equation is satisfied, however, then q could still be prime or composite.

What about trying different values of a? Yes, we can do that, and the chances are that if q is actually composite, sooner or later we will find a value of a that proves this. Unfortunately—yet again—there is a snag. Some numbers are composite and yet they satisfy Fermat's criterion for every value of a without exception. These are the **Carmichael numbers**.

Euler generalized Fermat's Little Theorem (in 1760), which only applies to prime numbers. He proved that if n is composite, and if a and n are coprime, then

$$a^{\phi(n)} \equiv 1 \ (\text{mod } n)$$

Fermat's Little Theorem implies that if p is prime then p divides $2^p - 2$. The converse claim, that if n divides $2^n - 2$ then n must be prime, has been attributed, quite wrongly, to the ancient Chinese, due to a mistranslation from the Chinese classic *The Nine Chapters of the Mathematical Art*.

The German philosopher and mathematician Leibniz, who invented the calculus at the same time as Newton, did believe the converse was true, understandably, because the first exception does not occur until $n = 341$.

See also AKS algorithm for primality testing; primality testing; Euler's totient (phi) function

Fermat quotient

By Fermat's Little Theorem, the quotient $(a^{p-1} - 1)/p$ is an integer if p is prime and a and p are coprime. It is usually denoted by $q_p(a)$ and behaves a bit like a logarithm:

If p does not divide a or b, then $q_p(ab) = q_p(a) + q_p(b)$

Fermat and primes of the form $x^2 + y^2$

Fermat also discovered one of the most beautiful and extraordinary properties in the theory of numbers, which we described in the introduction. It connects the very regular sequence of perfect squares with the apparently irregular primes.

Bachet de Méziriac, whose edition of **Diophantus** Fermat read and annotated, had claimed that "almost all" primes of the form $4n + 1$ are the sum of two integral squares, and Albert Girard stated that *all* primes $4n + 1$ have this property. Fermat proved that it was so.

If p is the prime 2 or a prime of the form $4n + 1$, greater than 3, then it is the sum of two integral squares in a unique way. The first few examples are:

$$2 \ = 1^2 + 1^2$$
$$5 \ = 1^2 + 2^2$$
$$13 = 2^2 + 3^2$$
$$17 = 1^2 + 4^2$$
$$29 = 2^2 + 5^2$$
$$37 = 1^2 + 6^2$$
$$41 = 4^2 + 5^2$$
$$53 = 2^2 + 7^2$$

It follows that if a number cannot be written as the sum of two integral squares, or can be so written in more than one way, then it is not prime.

Euler established in 1738 the condition that an integer is the sum of two squares: its prime factors of the form $4n + 3$ must all occur to an even power:

$$1^2 + 3^2 = 10 = 2 \cdot 5$$
$$1^2 + 5^2 = 26 = 2 \cdot 13$$
$$2^2 + 4^2 = 20 = 2^2 \cdot 5$$
$$3^2 + 4^2 = 25 = 5^2$$
$$3^2 + 5^2 = 34 = 2 \cdot 17$$
$$3^2 + 6^2 = 45 = 3^2 \cdot 5$$

If n is not a prime there is a formula for the number of ways, $r(n)$, in which it is the sum of two integral squares:

$$r(n) = 4(d_1(n) - d_3(n))$$

where $d_1(n)$ is the number of divisors of the form $4n + 1$ and $d_3(n)$ is the number of divisors of the form $4n + 3$. The function $r(n)$ counts "trivial" variations, so that $1 = (\pm 1)^2 + 0^2 = 0^2 + (\pm 1)^2$ and $r(1) = 4$.

This implies that $d_3(n)$ can never exceed $d_1(n)$. If this seems surprising, notice that we are talking about all the *divisors* of a number, not merely its prime factors. So if our number is $3689 = 7 \cdot 17 \cdot 31$, then although 7 and 31 are of the form $4n + 3$ and only 17 is a $4n + 1$, when we write down all its divisors, they are:

$$1 \quad \underline{7} \quad 17 \quad \underline{31} \quad \underline{119} \quad 217 \quad \underline{527} \quad 3689$$

and $d_1(n) = d_3(n)$ because 1 counts as a divisor (though not as a prime) and the product of 7 and $31 = 217$ is of the form $4n + 1$.

The remaining primes, which apart from 2 must be of the form $4n + 3$, cannot be the sum of two squares.

Fermat correctly claimed that a number is the sum of three squares unless it is of the form $4^n(8n + 7)$.

He also knew that all primes of the form $8n + 1$ or $8n + 3$ are of the form $x^2 + 2y^2$, all primes of the form $3n + 1$ can be expressed as $x^2 + 3y^2$, and if a number is the product of two primes of the form $20n + 3$ or $20n + 7$, then it is of the form $x^2 + 5y^2$. He also noticed that there were no matching properties for some forms such as $x^2 + 5y^2$.

There are other relationships of this kind, of which the simplest is: any prime p of the form $3n + 1$ is also of the form $x^2 + xy + y^2$. For example, $7 = 1^2 + 1 \cdot 1 + 2^2$; $13 = 1^2 + 1 \cdot 3 + 3^2$; $19 = 2^2 + 2 \cdot 3 + 3^2$.

Similarly, any prime of the form $6n + 1$ can be written as $a^2 + 3b^2$, and every prime of the form $8n + 1$ is also of the form $x^2 - 2y^2$. Since $x^2 - 2y^2 = (3x + 4y)^2 - 2(2x + 3y)^2$, it follows that there is an infinity of such representations. For example,

$$17 = 5^2 - 2 \cdot 2^2 = 23^2 - 2 \cdot 16^2 = 133^2 - 2 \cdot 94^2 = \ldots$$

Fermat's conjecture, Fermat numbers, and Fermat primes

Fermat also studied the sequence of numbers $2^{2^n} + 1$. They are now known as Fermat numbers. The first few are:

n	0	1	2	3	4	5
F_n	3	5	17	257	65537	composite

Fermat knew that the first five are all prime, and he famously conjectured that they all are: which turned out to be an equally famous mistake. Euler proved in 1732 that 641 is a factor of F_5, using the fact that he had already proved that any prime factor of F_n, if $n > 2$, is of the form $k \cdot 2^{n+1} + 1$. (**Lucas** later showed that k must be even, so this becomes $k \cdot 2^{n+2} + 1$.) So all its prime factors are of the form $128k + 1$, and the two smallest primes of this form are 257 and 641. As it happens, $257 = F_3$ and every pair of Fermat numbers is coprime, because of this identity:

$$F_0 F_1 F_2 \ldots F_{n-1} + 2 = F_n$$

Incidentally, this also proves that the number of primes is infinite!

Euler then spotted that $16 = 641 - 5^4$ and that $641 - 1 = 5 \times 2^7$, and so proved that 641 divides F_5:

$$F_5 = 2^{32} + 1 = (16)2^{28} + 1 = (641 - 5^4)2^{28} + 1 = 641m - (5 \cdot 2^7)^4 + 1 = 641m - (641 - 1)^4 + 1 = 641m - 1 + 1 = 641m$$

The simplest **primality test** for Fermat numbers is **Pepin's test**.

It has been conjectured that the number of Fermat primes is finite: but also that it is infinite.

The famous unresolved cases used to be $(2^{148} + 1)/17$ (identified in 1951 by A. Ferrier by using a hand-computing method), and $180 \cdot (2^{127} - 1)^2 + 1$ (settled in 1951 by Miller and Wheeler). Today all the Fermat primes up to 2 billion digits are known.

Another unresolved conjecture is that all Fermat numbers are **squarefree**. It is known that if F_n has a squared prime factor p^2, then p is a **Wieferich prime**, meaning that $2^{p-1} \equiv 1 \pmod{p^2}$, and such primes are very rare.

Fermat factorization, from F_5 to F_{30}

Fermat numbers are known to be composite from $n = 5$ to 30, but only F_5 to F_{11} have been completely factorized:

Fermat number	year	prover
F_5	1732	Euler found two factors: 641 and 6700417.
F_6	1856	Lucas proved that it was composite without showing the factors—the first such achievement.
F_6	1880	Landry (aged eighty) completely factorized it: $2^{64} + 1 = 274,177 \cdot 67,280,421,310,721$.
F_7	1905	Morehead and Western independently showed that it is composite.

Continued on next page

F_7	1975	Brillhart and Morrison factorized it into these two prime factors: $2^{128} + 1 = 34028236692093$ $846346337460743176821457 = 596495891274$ $97217 \cdot 5704689200685129054721$.
F_8	1981	Brent and Pollard completely factorized it.
F_9	1993	A. K. Lenstra, H. W. Lenstra Jr., M. S. Manasse and J. M. Pollard completely factorized it.
F_{10}	1999	Richard P. Brent completely factorized it; four factors of 8, 10, 40, and 252 decimal digits.
F_{11}	1899	Cunningham proved it to be composite.
F_{11}	1988	Brent and Morain completely factorized it.

These mathematicians proved that F_{12} to F_{30} are composite:

Fermat number	year	prover
F_{12}	1877	I. M. Pervushin and E. Lucas
F_{13}	1960	G. A. Paxson
F_{14}	1961	A. Hurwitz and J. L. Selfridge
F_{15}	1925	M. Kraitchik
F_{16}	1953	J. L. Selfridge
F_{17}	1980	G. B. Gostin
F_{18}	1903	A. E. Western
F_{19}	1962	H. Riesel
F_{20}	1987	Buell and Young
F_{21}	1963	C. P. Wrathall
F_{22}	1993	Carandall, Doenias, Norrie, and Young
F_{23}	1878	I. M. Pervushin
F_{24}	1999	Crandall, Mayer, and Papadopoulos
F_{25}	1963	C. P. Wrathall
F_{26}	1963	C. P. Wrathall
F_{27}	1963	C. P. Wrathall
F_{28}	1997	Taura
F_{29}	1980	G. B. Gostin and P. B. McLaughlin
F_{30}	1963	C. P. Wrathall

To date, 217 Fermat numbers are known to be composite. Wilfrid Keller keeps an extremely detailed account of Fermat numbers, their factors, factoring status, newly found factors, and so on, at www .prothsearch.net/fermat.html. His latest News Flash (as of November 5, 2004) is:

News Flash!

On October 5, 2004, Payam Samidoost discovered this new factor of a Fermat number: $89 \cdot 2^{472099} + 1$ divides F_{472097}.

Keller's two previous flashes announced that a new largest known composite Fermat number had been discovered by John Cosgrave and his Proth-Gallot Group at St. Patrick's College (Dublin, Ireland) on October 10, 2003: $3 \cdot 2^{2478785} + 1$ divides $F_{2478782}$; and that on November 1, 2003, Craig Kitchen found another new factor of a Fermat number: $1054057 \cdot 2^{8300} + 1$ divides F_{8298}, bringing to 250 the total number of then known prime factors.

Generalized Fermat numbers

These are numbers of the form, $a^{2^n} + b^{2^n}$. The Generalized Fermat Prime Search is organized by Phil Carmody using a new and powerful test, which enables Generalized Fermat primes to be found as quickly as **Mersenne primes** of the same size. If you fancy joining in, here are some of the records you have to beat!

- Yves Gallot on January 8, 2004 discovered the fifth Generalized Fermat prime of the form $b^{2^{17}} + 1$, which is $572186 \cdot 2^{17}$, with 754,652 digits.
- On August 22, 2003, Daniel Heuer discovered the largest known Generalized Fermat prime: $1176694 \cdot 2^{17} + 1$. This has 795,695 digits and became the fifth largest known prime.
- A month later, on September 22, 2003, Daniel Heuer beat his own record and found the new largest known Generalized Fermat prime: $1372930 \cdot 2^{17} + 1$. This has 804,474 digits and is now the fifth largest known prime.
- The four largest primes found by the Generalized Fermat Prime Search, all discovered in 2003–4, have more than 600,000 digits.
- The factor 641, which Euler proved divides F_5, also divides all the Generalized Fermat numbers of the form $(2^a 5^b)^{2^5} + 1$, where one of a and b is odd and the other even. (Yves Gallot; see http://perso.wanadoo.fr/yves.gallot/primes/index.html)

Fermat's Last Theorem

I think it's very important that people are encouraged to work on very hard problems. The tendency today is to work on short and immediate problems.

—attributed to Andrew Wiles

Fermat had his own copy of the works of **Diophantus** as translated by Bachet de Méziriac in 1621. In the margin he wrote, in 1637, originally in Latin:

> To divide a cube into two other cubes, a fourth power or in general any power whatever into two powers of the same denomination above the second is impossible, and I have assuredly found an admirable proof of this, but the margin is too narrow to contain it.

To which he added, "And perhaps, posterity will thank me for having shown it that the ancients did not know everything." (Burton 1976)

Fermat's only published proof, in 1659, was of Fermat's Last Theorem for the case $n = 4$, in which he used his famous method of *infinite descent*, showing that if one solution existed, then another, smaller, solution must also exist, and then a smaller solution still . . . and so on, reducing the problem to absurdity.

Few believe Fermat had such a proof, and Andrew Wiles found the first accepted proof in 1994, some 350 years later, but only after eight years of working secretly, during which he published occasional minor papers to put colleagues off the scent and allay any suspicions as to what he was really up to.

Erdös, the great collaborator, not surprisingly objected to Wiles's secrecy and claimed that the problem would have been solved sooner if Wiles had been open: maybe yes, but maybe not by Wiles! Mathematicians are as competitive as anyone—and Erdös himself had an unfortunate experience with Atle Selberg over their elementary proof of the **prime number theorem**, for which Selberg but not Erdös won the Fields Medal in 1950. (Hoffman 1998, 183)

The philosopher and logician W. V. Quine pointed out that Fermat's Last Theorem is equivalent to this claim about sorting objects into bins:

Suppose that there are z bins in total, and that x bins are not painted red, and y bins are not painted blue. The remaining bins are uncolored. You have n objects to sort into the bins.

Then Fermat's Last Theorem is equivalent to the statement that: "The number of ways of sorting them that shun both colors is equal to the number of ways that shun neither." (Quine 1988)

the first case of Fermat's Last Theorem

Prime numbers have appeared more than once in attempts to settle Fermat's Last Theorem, by Kummer, **Sophie Germain**, and **Wieferich**.

The first case is when x, y, z, and p are coprime. The second case is when p divides at least one of x, y, and z. The distinction was first made by Sophie Germain in 1832, who proved that for a prime p, $x^p + y^p = z^p$ has no solution if $2p + 1$ is also prime.

Another test for the first case, by Mirimanoff in 1910, says that in the first case, if there are solutions, then $m^{p-1} \equiv 1 \pmod{p^2}$ is true for $m = 2$ and $m = 3$. No such primes are known.

Ernst Eduard Kummer realized in 1843 that current attempts to prove Fermat's Last Theorem did not work because **Gaussian integers** might be factorized in several different ways: their factorization was not unique. So he invented ideal numbers, one of the many developments in mathematics that have been prompted by attempts to solve FLT.

Kummer proved Fermat's Last Theorem for all *regular* primes, where an *irregular* prime is an odd prime p that divides the numerator of one of the **Bernoulli numbers** B_{2n} with $2n + 1 < p$. The first irregular prime to appear in the sequence of numerators is 691 at B_{12}.

Kummer proved Fermat's Last Theorem for all odd prime exponents less than 165, except for these eight irregular primes: 37, 59, 67, 101, 103, 131, 149, and 157.

The sequence of regular primes starts, 3, 5, 7, 11, 13, 17, 19, 23, 29, 31, 41, 43, 47, 53, 61, 71, 73, 79, 83, 89, 97, . . . but it is not known if there is an infinity of them.

The proportion of regular primes among the primes less than 10^6 is $47627/78497 = 0.6067 \ldots$ which fits the conjecture that the density tends to $e^{-1/2} = 0.6065 \ldots$ (e is the base of natural logarithm).

Jensen proved in 1915 that there is an infinity of irregular primes, and all the irregular primes up to 12,000,000 have now been calculated.

Wall-Sun-Sun primes

D. D. Wall, Z. H. Sun, and Z. W. Sun proved in 1992 that if the first case of **Fermat's Last Theorem** is false for p, then p^2 divides $(p - (p|5))$th **Fibonacci number**, where $p|5$ is the Legendre symbol used to express the law of **quadratic reciprocity**. (This notation is *nothing* to do with $p|5$ meaning "p divides 5 exactly"!)

We now know that Fermat's Last Theorem is never false, but it still remains an unknown question whether any Wall-Sun-Sun primes exist.

See also Beal's conjecture; Bernoulli numbers; factorization; Fermat-Catalan equation and conjecture; pseudoprimes; Sophie Germain primes; Wieferich primes

Fermat-Catalan equation and conjecture

Catalan's conjecture and **Fermat's Last Theorem** are both special cases of the Fermat-Catalan equation:

$$x^p + y^q = z^r$$

Here x, y, and z are positive, coprime integers and the exponents are all primes with $1/p + 1/q + 1/r \leq 1$.

The Fermat-Catalan conjecture is that there are only finitely many solutions to this system. These solutions include:

$$
\begin{aligned}
1 + 2^3 &= 3^2 \\
17^3 + 2^7 &= 71^2 \\
7^2 + 2^5 &= 3^4 \\
11^4 + 3^5 &= 122^2 \\
13^2 + 7^3 &= 2^9
\end{aligned}
$$

More recently, F. Beukers and D. Zagier have found these solutions, making a total of ten known solutions:

$$
\begin{aligned}
1549034^2 + 33^8 &= 15613^3 \\
2213459^2 + 1414^3 &= 65^7 \\
15312283^2 + 9262^3 &= 113^7 \\
76271^3 + 17^7 &= 21063928^2 \\
96222^3 + 43^8 &= 30042907^2
\end{aligned}
$$

(Crandall and Pomerance 2001, 383)

See Beal's conjecture

Fibonacci numbers

Leonardo Pisano (c. 1175–1250), nicknamed Fibonacci, in his *Liber Abaci* brought the Hindu-Arabic numeral system to Western Europe in 1202. He also listed the primes from 10 to 100 and pointed out that to check whether a number was prime you only needed to divide it by the primes less than its square root. He also included problems similar to **Diophantus**, such as how to find a square that remains a square when 5 is added or subtracted. His answer, in fractions, is,

$$(41/12)^2 + 5 = (49/12)^2$$
$$(41/12)^2 - 5 = (31/12)^2$$

Fibonacci is now remembered for this problem:

> A certain man put a pair of rabbits in a place surrounded on all sides by a wall. How many pairs of rabbits can be produced from that pair in a year if it is supposed that every month each pair begets a new pair which from the second month on becomes productive?

Assuming that the rabbits are immortal, this is the number of new pairs each month:

$$1 \quad 1 \quad 2 \quad 3 \quad 5 \quad 8 \quad 13 \quad 21 \quad 34 \quad 55 \quad 89 \ldots$$

Each term is the sum of the previous two terms.

This extraordinary sequence has so many properties that there is a journal, *The Fibonacci Quarterly*, devoted entirely to it. For example, the ratio of successive terms, F_{n+1}/F_n tends to the Golden Ratio, 1.618 . . .

To date, $F_n{}^*$ is known to be prime for n = 3, 4, 5, 7, 11, 13, 17, 23, 29, 43, 47, 83, 131, 137, 359, 431, 433, 449, 509, 569, 571, 2971, 4723, 5387, 9311, 9677, 14431, 25561, 30757, 35999, and 81839. All values of n up to 100,000 have been tested by Harvey Dubner and Wilfrid Keller.

It isn't known if there are an infinite number of Fibonacci primes. However, it is known that consecutive pairs are coprime, because of the identity,

$$F^2{}_{n+1} = F_n F_{n+2} + (-1)^n$$

---

*For simplicity and memorability, we are using F_n for both the nth Fermat number and the nth Fibonacci number. This should not cause confusion as they never seem to occur together!

There are many other formulae connecting the Fibonacci numbers, such as:

$$F^2_n + F^2_{n+1} = F_{2n+1}$$
$$F^2_n = F_{n-1}F_{n+1} - (-1)^n \quad \text{provided } n > 1$$
$$F^2_n = F_{n-2}F_{n+2} + (-1)^n \quad \text{provided } n > 2$$
$$F^2_n = F_{n-3}F_{n+3} - 2^2(-1)^n \quad \text{provided } n > 3$$
$$F^2_n = F_{n-4}F_{n+4} + 3^2(-1)^n \quad \text{provided } n > 4$$
$$F^2_n = F_{n-5}F_{n+5} - 5^2(-1)^n \quad \text{provided } n > 5$$

and so on. The coefficients of $(-1)^n$ are the Fibonacci squares.

These formulae show that F_{2n} is never adjacent to a prime, except possibly when $n = 3$.

$$F_{2n} + (-1)^n = F_{n-1}(F_{n+2} + F_n)$$
$$F_{2n} - (-1)^n = F_{n+1}(F_n + F_{n-2})$$

There are similar formulae for F_{2n+1}.

Vernon Hoggatt Jr. and Marjorie Bicknell-Johnson proved much more: the *smaller* neighbor of every power of a Fibonacci number is composite, with the exception of $F_3^2 = 4$.

The *larger* neighbors of all powers of F_n are also composite, unless the power itself is a power of 2 and n is a multiple of 3, in which case it may be prime. For example, $F_9^4 + 1 = 1336337$ and $F_{15}^8 + 1$ are both prime. (Hoggatt and Bicknell-Johnson 1977)

divisibility properties

A basic fact is that F_n divides F_{mn}, so if F_q is prime, then q must be prime, with the sole exception of $q = 4$, since $F_2 = 1$. However, this is a necessary but not sufficient condition. Thus $F_{19} = 4181 = 37 \cdot 113$, and $F_{37} = 73 \cdot 149 \cdot 2221$.

The greatest common divisor of two Fibonacci numbers, F_n and F_m, is always another Fibonacci number whose index is the greatest common divisor of n and m.

A related theorem is that if and only if $n \mid m$ then $F_n \mid F_m$, so you can study the divisibility of the Fibonacci numbers by studying their index numbers, and draw conclusions such as:

$2 \mid F_n$ if and only if $3 \mid n$
$3 \mid F_n$ if and only if $4 \mid n$
$5 \mid F_n$ if and only if $5 \mid n$ and so on . . .

We can also prove that there is at least one Fibonacci number divisible by any given number n: in fact, it can be found among the first n^2 numbers in the sequence. It then follows that there are an infinity of such numbers. In particular, for every *prime* number, p, there is a Fibonacci term that is divisible by p, and it occurs among the first $p + 1$ terms.

Choose $n + 1$ Fibonacci numbers from the set $F_1, F_2, F_3, \ldots, F_{2n}$: then one of the chosen numbers divides another, exactly. (Weinstein 1966)

If p is a prime greater than 7 *and* $p \equiv 2 \pmod 5$ or $p \equiv 4 \pmod 5$ *and* $2p - 1$ is also prime (a condition reminiscent of the **Sophie Germain prime** condition), then $(2p - 1) \mid F_p$, which is therefore composite.

Mihàly Bencze proved an elegant divisibility property for what might be called a very generalized Fibonacci sequence. In the sequence, the first four values are given, and the rule then is that $B(n + 4) = B(n + 1) + B(n)$:

n	0	1	2	3	4	5	6	7	8	9
$B(n)$	4	0	0	3	4	0	3	7	4	3
n	10	11	12	13	14	15	16	17	18	19
$B(n)$	10	11	7	13	21	18	20	34	39	38

$B(n)$ is always divisible by n, when n is prime. (Bencze 1998)

If p is prime, then F_p may be prime or composite, but if it is composite, then its factors will never have appeared earlier in the Fibonacci sequence as factors of any term.

Fibonacci curiosities

- The largest known Fibonacci prime is F_{81839}.
- For every n, it is possible to find n consecutive composite Fibonacci numbers. For $n \geq 4$, $F_n + 1$ is always composite.

- Every positive can be written as a sum of distinct Fibonacci numbers.
- The prime number 17 is the only prime that is the average of two consecutive Fibonacci numbers. (Honaker: Caldwell, *Prime Pages*)
- If p is prime, then F_p^n is square-free, unless $p = 5$, in which case $F_5^n = 5^n m$ where m is square-free.
- F_1, F_2, $F_6 = 8$ and $F_{12} = 144$ are the only Fibonacci numbers that are powers.
- If F_n is prime, then n is prime also, but the converse is occasionally false. The smallest counterexample is $F_{19} = 4181 = 37 \cdot 113$.
- The only Fibonacci perfect cubes are 1 and 8.
- The only numbers that are simultaneously Fibonacci and **Lucas** are 1 and 3.
- A. W. F. Edwards defined an *infinite coprime sequence* as one in which any pair of terms are coprime. A simple set of such sequences is given by: $u_n = u_{n-1}^2 - u_{n-1} + 1$, for $u_1 > 1$. For example, if $u_1 = 2$, the sequence starts 1, 3, 7, 43, 1807, 3263443, . . . (Edwards 1964)
- **Mersenne numbers** satisfy a Fibonacci-like equation: $M_{n+1} = 3M_n - 2M_{n-1}$

Édouard Lucas and the Fibonacci numbers

Lucas factorized all the first sixty Fibonacci numbers, and first noted that if $n \mid m$, then $F_n \mid F_m$. He then observed a remarkable fact about their *primitive factors*, meaning those factors that divide F_n but do not divide any smaller Fibonacci number. A primitive factor of F_n is congruent to $\pm 1 \pmod{n}$, with the exception $n = 5$.

n	F_n	factors	primitive factors (mod n)
1	1		
2	1		
3	2	2	$2 = 3 - 1$
4	3	3	$3 = 4 - 1$
5	5	5	(excepted from rule)
6	8	2^3	
7	13	13	$13 = 2 \cdot 7 - 1$
8	21	$3 \cdot 7$	$7 = 8 - 1$
9	34	$2 \cdot 17$	$17 = 2 \cdot 9 - 1$

10	55	$5 \cdot 11$	$11 = 10 + 1$
11	89	**89**	$89 = 8 \cdot 11 + 1$
12	144	$2^4 \cdot 3^2$	
13	233	**233**	$233 = 13 \cdot 18 - 1$
14	377	$13 \cdot 29$	$29 = 14 \cdot 2 + 1$
15	610	$2 \cdot 5 \cdot 61$	$61 = 15 \cdot 4 + 1$
16	987	$3 \cdot 7 \cdot 47$	$47 = 16 \cdot 3 - 1$
17	1597	**1597**	$1597 = 17 \cdot 94 - 1$
18	2584	$2^3 \cdot 17 \cdot 19$	$19 = 18 - 1$
19	4181	$37 \cdot 113$	$37 = 19 \cdot 2 - 1$ and $113 = 19 \cdot 6 - 1$
20	6765	$3 \cdot 5 \cdot 11 \cdot 41$	$41 = 20 \cdot 2 + 1$

Based on such observations—he had no proof, **Carmichael** proved the result much later in 1913—Lucas proposed as a theorem that:

If $n \equiv \pm 3 \pmod{10}$ and n is a primitive divisor of $F(n + 1)$, then n is prime. If $n \equiv \pm 1 \pmod{10}$ and n is a primitive divisor of $F(n - 1)$, then n is prime.

On the basis of this theorem, and with an estimated 170 to 300 hours of work, Lucas drew a remarkable conclusion: that the enormous number $2^{127} - 1$ is prime. As Lucas observed, his theorem "allows us to determine whether a number is prime or composite without making use of a table of prime numbers," and so he inaugurated the modern era of primality testing.

See also Fibonacci composite sequences

Fibonacci composite sequences

Although it's not known if there is an infinity of Fibonacci primes, there are *generalized Fibonacci sequences* in which each term is the sum of the previous two terms, which contain no prime numbers: every term is composite.

In 1964 R. L. Graham believed he had constructed an example of such a sequence, for which:

$$G(1) = 1786\ 77270\ 19288\ 02632\ 26871\ 51304\ 55793$$
$$G(2) = 1059\ 68322\ 50539\ 15111\ 05816\ 51416\ 86995$$

The sequence then continues with $G(3) = G(1) + G(2)$, and so on.

But he made a mistake that was corrected by Donald Knuth (1990), who gave these correct starting numbers:

$G(1) = 331\ 63563\ 59982\ 74737\ 47220\ 06564\ 30763$
$G(2) = 1510\ 02891\ 10884\ 01971\ 18959\ 03054\ 98785$

Knuth then produced his own smaller starting pair:

$G(1) = 49\ 46343\ 57432\ 05655$
$G(2) = 62\ 63828\ 00042\ 39857$

(Graham 1964) (Knuth 1990)

formulae for primes

A formula, a magic rule, for producing primes! How delightful! Unfortunately, no *simple* and *useful* formula exists, though no one has ever proved that a formula is impossible. We do, however, have some good tries, some of them merely amusing, some of them amazing, and some of them deep.

The term "formula" itself is ambiguous. It could mean a formula that produces all the primes *pn* in order, as a function of *n*. So you simply apply the formula to find the *n*th prime. An exact formula for $\pi(n)$ would be effectively as good.

A much weaker demand is that the formula produces only prime numbers, never composite. This is what Fermat thought, mistakenly, of his **Fermat numbers**, and many other conjectures that a certain formula produced only primes have also been shot down, though there are many formulae that produce primes *for a while* and then fail, illustrating Richard Guy's **strong law of small numbers**.

Conversely, the expression $n^6 + 1091$ is *composite* for $n = 1$ to 3905 (Shanks), and $2^{2^n} - 5 - 2^a$ is composite for all $n > 2$ and $a > 2$. (Crocker 1961)

An even weaker interpretation is a function that produces surprisingly many primes. The *linear* polynomial $30n - 13$ produces 411 primes among the first 1,000 values of *n*. (Caldwell, *Prime Pages*) Better-known examples are **Euler's prime polynomial** and its relatives.

The very word "formula" is also unclear. In Hollywood movies, any mathematical formula, stolen by a fiendish spy or created by a mad professor, is likely to be a polynomial, with maybe a few square root signs or integrals thrown in. We can allow more functions than this: for example, factorial, subfactorial, the integral part of a non-integer, and many others.

Conway's prime-producing machine is extraordinarily impressive at first glance, but it turns out to be a cunningly disguised sieve.

No polynomial can produce only primes, but in solving **Hilbert's 10th problem**, Matijasevic showed that there are polynomials in many variables whose positive values are the prime numbers—though most of their values are negative and usually not prime.

If $y \equiv x \pmod{n}$ is an allowable function, this function generates all the primes, if n is any non-negative integer:

$$f(n) \equiv 2 + (2(n!) \bmod (n + 1))$$

If $n + 1$ is prime, then by **Wilson's theorem**

$$n! \equiv -1 \pmod{n + 1}$$
and so $\qquad 2(n!) \equiv -2 \pmod{n + 1}$
and so $\qquad f(n) = n + 1.$

But if $n + 1$ is prime, then its prime factors will be less than n and so $2(n!) \equiv 0 \pmod{n + 1}$, and $f(n) = 2$.

C. P. Willans established a formula to test whether any number, x, is prime. It is

$$f(x, y) = \tfrac{1}{2}(y - 1) \lfloor \, |B^2(x,y) - 1| - (B^2(x, y) - 1) \rfloor + 2$$

where $B(x,y) = x(y + 1) - (y! + 1)$ and $\lfloor x \rfloor$ is the floor function, equal to the greatest integer $\leq x$.

This function also generates each odd primes once, and otherwise equals 2. (Honsberger 1976, 33) For example,

$$F(1,2) \quad = 3$$
$$F(5,4) \quad = 5$$
$$F(103,6) = 7$$

No other prime is generated for values of x, y less than 1000.

Willans then used this function to derive a complex and typically useless formula for the nth prime number. (Willans 1964, 413–15)

Here is another trick: define the number s to be the decimal in which each prime p is followed by p zeros:

$$s = 0.200300050000070000000110000000000013 \ldots$$

Then the formula

$$p_n = \lfloor 10^{n^2}s \rfloor - 10^{2n-1} \lfloor 10 \exp{(n-1)^2 s} \rfloor$$

recovers them, but tells you nothing you didn't already know *unless* there were a means of calculating s independently, which no one expects.

G. H. Hardy managed to find an exact formula for $\vartheta(x)$, the largest prime factor of a number x. Without making the slightest attempt to explain it, here it is!

$$\theta(x) = \lim_{r=\infty} \lim_{m=\infty} \lim_{n=\infty} \sum_{v=0}^{m} [1 - (\cos{\{(v!)^{r\pi}/x\}})^{2n}]$$

(Dudley 1983)

See also Euler's quadratic; Mills' theorem

Fortunate numbers and Fortune's conjecture

Take the sequence of **primorials**, the products of the primes in sequence:

		2	6	30	210	2310	30030	...
add 1		3	7	31	211	2311	30031	...
take the next prime	5	11	37	223	2333	30047		...
take the difference	2	4	6	12	22	16		...
add 1		3	5	7	13	23	17	...

The lengths some people will go to produce prime numbers! Reo Fortune, who was an anthropologist and once married to Margaret Mead, thought up this one and conjectured, naturally, that all the *Fortunate numbers* are prime.

The sequence goes: 3, 5, 7, 13, 23, 17, 19, 23, 37, 61, 67, 61, 71, 47, 107, 59, 61, 109, 89, 103, 79, 151, 197, 101, 103, 233, 223, 127, 223, 191, 163, 229, 643, ...

The conjecture may well be true, since the nth Fortunate number is not divisible by any of the first n primes, and yet is relatively small. Thus the fourth Fortunate number is not divisible by 2, 3, 5, or 7, and is only 13, so it must be prime (even if we didn't know that already!).

Indeed, we could replace "the next prime" by "the next prime-but-one" and still expect to get primes:

		2	6	30	210	2310	. . .
add 1		3	7	31	211	2311	. . .
take the next prime	7	13	41	227	23339	. . .	
take the difference	4	6	10	16	28	. . .	
add 1		5	7	11	17	29	. . .

The conjecture could only fail if there is a sufficiently large gap between two primes, so that "the next prime," on at least one occasion, is very large indeed, and this is thought unlikely. (Golomb 1981)

gaps between primes and composite runs

The gap $g(n)$ between the primes $p(n)$ and $p(n + 1)$ is usually defined to be the number of composite numbers between them, so $g(n)$ is 1 less than the difference.

Since the primes slowly become scarcer and scarcer, $g(n)$ must increase on average. In fact, we can find gaps eventually as large as we please. However, there are an infinite number of n for which $g(n + 1) < g(n)$.

We say that $g(n)$ is a maximal gap if $g(m) < g(n)$ for all $m < n$. A maximal gap is a record-breaking gap, the first occurence of a gap that large, up to that point.

The sequence of maximal gaps starts with $n = 1, 2, 4, 9, 24, 30, 99,$ 154, 189, 217, 1183, 1831, 2225, 3385, 14357, 30802, . . .

Cramér's conjecture is that the largest gap between primes round about n is always less than $(\log n)^2$. If Cramér's conjecture is true, then so is **Fortune's conjecture**, at least for large values of n.

If $n > 48$, then there is a prime between n and $9n/8$, inclusive.

If $n > 117$, then the interval n to $4n/3$ inclusive contains a prime number of each of the forms $4n + 1$, $4n + 3$, $6n + 1$, and $6n + 5$.

These theorems require difficult proofs. In contrast, computers have been used to discover many curious statistics, such as:

- Among the first six million primes (up to 104,395,289), the largest difference between consecutive primes is 220, and the smallest difference that does not appear in this range is 186. (Gruenberger and Armerding)

- The first pair of consecutive primes differing by 100 are 396,733 and 396,833.
- The first gap of at least 1,000 is the surprisingly large gap of 1,132 following the prime 1693182318746371, discovered by Thomas Nicely and Bertil Nyman.
- Harvey Dubner has found a gap of at least 50,206 near $3 \cdot 10^{1883}$.

Gauss, Johann Carl Friedrich (1777–1855)

Mathematics is the queen of the sciences and number theory is the queen of mathematics.

—Gauss

Gauss was a genius who contributed to almost all fields of pure and applied mathematics, especially calculus, geometry, algebra, probability, geodesy, and of course number theory, as well as—like Newton—astronomy and optics.

As a child he was a calculating prodigy, but unlike most prodigies he retained this useful ability all his life. It contributed to the pattern of his work, which combined extensive empirical investigations with deep insight and an emphasis on proof, not merely to confirm what he was already convinced was true, but as a source of illumination.

He published six proofs of the law of **quadratic reciprocity** (a seventh was found among his papers), each one adding to his—and other mathematicians'—understanding of this deep problem.

He showed at the age of nineteen that a regular n-sided polygon can be constructed with a ruler and compasses if and only if n is the product of a power of 2 and one or more distinct **Fermat primes**. Just as Euclid had ended his *Elements* with the construction of the do decahedron and icosahedron, Gauss ended his early masterpiece, *Disquisitiones Arithmeticae*, published in 1801 when he was twenty-four, with the solution to this problem and a list of possible numbers of sides up to the limit of 300.

In the same year he used novel methods to predict the position of the minor planet Ceres, which had been "lost," a sensational result after which he was famous as both a mathematician and an astronomer.

Gauss also proved that every number is the sum of three triangular numbers, a conclusion that Fermat had drawn but could not prove. Gauss reduced it to the problem of representing a number as the sum of three squares by pointing out that if,

$$M = \tfrac{1}{2}x(x + 1) + \tfrac{1}{2}y(y + 1) + \tfrac{1}{2}z(z + 1)$$

Then $\qquad 8M + 3 = (2x + 1)^2 + (2y + 1)^2 + (2z + 1)^2$

He then proved another of Fermat's claims, that any integer is the sum of at most four squares.

He refused to spend time on **Fermat's Last Theorem**, replying to the astronomer Olber's urging by saying, "I confess that Fermat's Theorem as an isolated proposition has very little interest for me, because I could easily lay down a multitude of such propositions, which one could neither prove nor dispose of." (Newman 1956)

Gauss and the distribution of primes

Gauss, in a letter to Encke on Christmas Eve 1849, wrote, "As a boy I considered the problem of how many primes there are up to a given point. From my computations, I determined that the density of primes around x, is about $1/\log(x)$." He claimed to have made many of his discoveries "through systematic, palpable experimentation." (Mackay 1994)

Given a few minutes' spare time he would calculate prime numbers. He is supposed during his lifetime to have calculated all the primes up to 3,000,000. However, he also employed another calculating prodigy, Zacharias Dase (1824–1861), who was no mathematician, to do calculations for him. Like many calculating prodigies, Dase toured widely giving exhibitions of his abilities. He calculated seven-figure tables of logarithms and the value of π to two hundred places, and made a factor table of the numbers between 7 and 10 million.

Gauss later conjectured on the basis of empirical evidence that $\pi(x)$, the number of primes less than x, is approximately equal to

$$Li(x) = \int_2^x \frac{dt}{\log t}$$

and that the number of primes up to n is about $n/(\log n)$.

This is the table that he sent to Encke, as corrected by D. N. Lehmer.

x	primes < x	$Li(x) = \int dt/\log t$	difference
500,000	41,538	41,606	68
1,000,000	78,498	78,628	130
1,500,000	114,155	114,263	108
2,000,000	148,933	149,055	122
2,500,000	183,072	183,245	173
3,000,000	216,816	216,971	155

Gaussian primes

Gaussian integers are of the form $a + bi$ where $i = \sqrt{-1}$. Like the natural numbers, Gaussian integers can be uniquely factorized. However, certain ordinary primes are composite when regarded as Gaussian integers with $b = 0$.

In fact, since all ordinary primes of the form $4n + 1$ can be expressed as the sum of two squares, such as

$$13 = 2^2 + 3^2$$

they are Gaussian composites, because

$$2^2 + 3^2 = (2 - 3i)(2 + 3i)$$

A Gaussian prime $a + bi$ fits these conditions: if a and b are non-zero, then $a + bi$ is a Gaussian prime if and only if $a^2 + b^2$ is an ordinary prime. If either a or $b = 0$, then the absolute value of the other must be an ordinary prime of the form $4n + 3$.

One measure of the "size" of a Gaussian integer, $a + bi$, is its *norm*, $\sqrt{a^2 + b^2}$. The sequence of Gaussian primes in the first quarter of the Argand diagram starting with the smallest norm is: $1 + i, 2 + i, 3, 3 + 2i, 4 + i, 5 + 2i, 7, 7 + 2i, 6 + 5i, 8 + 3i, 8 + 5i, 9 + 4i, \ldots$

If $a + bi$ is a Gaussian prime, then so are $a - bi$, $-a + bi$, and $-a - bi$, so the Gaussian primes are symmetrical about the origin when they are plotted on an Argand diagram. This is the pattern they make, for all a and b such that $\sqrt{a^2 + b^2} < 1000$, meaning that they lie within a circle with radius $10\sqrt{10}$.

This pattern has been used for tablecloths and tiling floors, as well as prompting questions such as: if the Gaussian primes are "stepping-stones" to infinity, what is the largest step you ever need to take? It is known that you need to cross gaps of at least 5. (Gethner, Wagon, and Wick 1998)

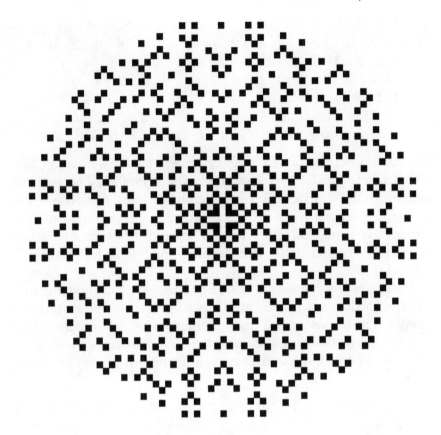

Gauss's circle problem

How many points are there on or inside a circle whose center is a lattice point? If the radius is R, then the total is,

$$r(0) + r(1) + r(2) + r(3) \ldots + r(R^2)$$

where $r(N)$ is the formula for the number of ways in which N is the sum of two integral squares. (*See* Fermat and primes of the form $x^2 + y^2$.)

The points on a square lattice that are visible from the origin are just the points whose coordinates are *coprime*: so their density is equal to the probability that two integers chosen at random are coprime, which is $6/\pi^2$.

See Fermat; prime number theorem

Gilbreath's conjecture

Make up a table of repeated absolute differences of the prime numbers in sequence, like this:

```
2  3  5  7  11  13  17  19  23  29  31  37  41  43  47  53  59  61  67  71
 1  2  2  4   2   4   2   4   6   2   6   4   2   4   6   6   2   6   4
  1  0  2  2   2   2   2   2   4   4   2   2   2   2   0   4   4   2
   1  2  0  0   0   0   0   2   0   2   0   0   0   2   4   0   2
    1  2  0  0   0   0   2   2   2   2   0   0   2   2   4   2
     1  2  0  0   0   2   0   0   0   2   0   2   0   2   2
      1  2  0  0   2   2   0   0   2   2   2   2   2   0
```

The pattern is pretty much irregular, except for the left-hand edge, which after the initial 2 seems to always be 1. Proth first noticed this pattern in 1878 and claimed, mistakenly, to have proved it. Later Gilbreath spotted it when he was a student at the University of California at Los Angeles in the late 1950s, and it is now called Gilbreath's conjecture. It has been verified for primes less than 10^{13}. (Odlyzko 1993)

One ingenious suggestion is that this property has nothing specifically to do with primes at all, but is true of any starting sequence that begins with 2 and continues with odd numbers that increase fairly slowly and don't have too great gaps. Here is an example, in which the initial row is a mixture of primes and composites:

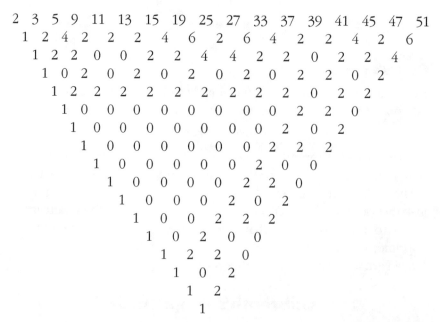

But why? No one knows.

GIMPS—Great Internet Mersenne Prime Search

GIMPS is the best-known example of **distributed computing**. In 1995 George Woltman collated all the known results on **Mersenne primes** and in January 1996 put them on the Web, together with a fast program for testing whether a Mersenne number is prime. Soon scores of experts and thousands of amateur enthusiasts were joining the hunt, including schoolteachers who have used the GIMPS phenomenon to motivate their students, who can run the free software themselves as a class project.

Today, the GIMPS uses Entropia's PrimeNet system, which performs 2 trillion calculations per second, every second, all day, while its tame computers are doing other tasks, even while their owners are asleep—nothing stops it!

The results are incredible: GIMPS spent thirteen thousand years of computer time to find the then record-breaking prime $2^{13,466,917} - 1$, discovered by Michael Cameron in 2001.

One GIMPS enthusiast, Nayan Hajratwala of Plymouth, Michigan, won $50,000 in 1999, one of the **Electronic Frontier Foundation**'s cooperative computing awards, for discovering the first million-digit prime, $2^{6,972,593} - 1$, which has 2,098,960 digits.

If you want to join the fun, more information and the free software are available at www.mersenne.org/prime. The state of play is recorded on the GIMPS status page, at www.mersenne.org/status.

All the Mersenne numbers less than $10^{1,000,000}$ have been tested at least once, most of them more than once. Both of the largest known primes were found by GIMPS members. On December 11, 2003, CNN headlined, "Student finds largest known prime number." A GIMPS member, Michael Shafer, a graduate student at Michigan State University, had found the latest world record prime on his Dell computer, one of more than two hundred thousand computers involved in the search, making between them 9 trillion calculations per second. It is $2^{20,996,011} - 1$, is 6,320,430 digits long, and would take hundreds of pages to print out.

Giuga's conjecture

According to **Fermat's Little Theorem**, if p is prime then each of the $p-1$ terms on the left-hand side is equal to 1 (mod p), so their sum equals -1 (mod p):

$$S_p = 1^{p-1} + 2^{p-1} + 3^{p-1} + \ldots + (p-1)^{p-1} \equiv -1 \; (\text{mod } p)$$

Giuga conjectured in 1950 that if this congruence is true, p must be prime.

He also proved that $S_n \equiv -1$ (mod n) if and only if for each prime divisor p of n, both $p-1$ and p divide $n/p - 1$, so if a counterexample exists it would be a special type of **Carmichael number**, which looks like a very limiting condition. Indeed, if an exception to Giuga's conjecture exists, it has at least 12,000 digits. This conjecture has been verified for $p \leq 10^{1700}$. (Ribenboim 1996)

Giuga numbers

If n is a composite number such that $p \mid n/p - 1$ for all prime divisors, p, of n, then n is a Giuga number. Giuga proved that n is a Giuga number if and only if the sum of the reciprocals of its prime divisors, less their product, is an integer. For example, the smallest Giuga number is 30 and,

$$\tfrac{1}{2} + \tfrac{1}{3} + \tfrac{1}{5} - \tfrac{1}{30} = 1$$

The sequence of Giuga numbers continues $858 = 2 \cdot 3 \cdot 11 \cdot 13$

for which $\tfrac{1}{2} + \tfrac{1}{3} + \tfrac{1}{11} + \tfrac{1}{13} - \tfrac{1}{858} = 1$
followed by $1722 = 2 \cdot 3 \cdot 7 \cdot 41$
for which $\tfrac{1}{2} + \tfrac{1}{3} + \tfrac{1}{7} + \tfrac{1}{41} - \tfrac{1}{1722} = 1$
66198 $= 2 \cdot 3 \cdot 11 \cdot 17 \cdot 59$
2214408306 $= 2 \cdot 3 \cdot 11 \cdot 23 \cdot 31 \cdot 47057$
24423128562 $= 2 \cdot 3 \cdot 7 \cdot 43 \cdot 3041 \cdot 4447$
. . .

Five Giuga numbers with seven or eight factors are known. It is not known if there are an infinity of Giuga numbers. (Borwein, Borwein, Borwein, and Girgensohn 1996)

Goldbach's conjecture

The famous problem of which I propose to speak tonight is probably as difficult as any of the unsolved problems of mathematics.

—*G. H. Hardy on Goldbach's conjecture*

Christian Goldbach (1690–1764) was born in Königsberg in Prussia but spent most of his life in Russia where he was appointed correspondence secretary of the Imperial Academy of Sciences at St. Petersburg. His social success interfered with his mathematical work but he corresponded with many of the leading mathematicians of Europe, and for more than thirty years with **Euler**, whom he prompted to investigate the fifth **Fermat number**, $2^{2^5} + 1$, which Euler proved was composite.

On June 7, 1742, Goldbach wrote a letter to Euler in which he speculated on how numbers might be represented as the sum of primes. In particular, he suggested that every number is the sum of three primes.

Goldbach's conjecture means today the strong claim that any even number (2 excepted) is the sum of two primes: it has never been proved.

The weaker conjecture, that every odd number can be written as a sum of three primes, is known to be true if another conjecture related to the **Riemann hypothesis** is true. It is true anyway for all but a finite but very large number of possible exceptions, all less than 10^{43000}.

The main conjecture remains extremely difficult, so the publishers Faber and Faber must have felt pretty safe when they announced as an advertising gimmick to promote their book *Uncle Petros and Goldbach's Conjecture* by Apostolos Doxiadis that they were offering £1 million to anyone who could prove Goldbach's conjecture between March 20, 2000, and March 20, 2002. They were right—the prize was not claimed.

Ironically, Goldbach himself would not have been allowed to enter—Item 10 of the Terms and Conditions restricted the challenge to "legal residents of the United Kingdom and the United States aged 18 or over"!

Although Goldbach's conjecture has not been proved, it is clear (!) that all even numbers but the smallest are the sum of two primes in an increasing number of ways:

$$14 = 3 + 11 = 7 + 7$$
$$16 = 3 + 13 = 5 + 11$$
$$18 = 5 + 13 = 7 + 11$$
$$20 = 3 + 17 = 7 + 13$$
$$22 = 3 + 19 = 5 + 17 = 11 + 11$$
$$24 = 5 + 19 = 7 + 17 = 11 + 13$$
$$26 = 3 + 23 = 7 + 19 = 13 + 13$$
$$28 = 5 + 23 = 11 + 17$$

. . .

The sequence of the smallest even numbers that are the sum of two primes in n ways starts: 6, 10, 22, 34, 48, 60, . . .

Goldbach's conjecture was verified in 1998 by Joerg Richstein up to $4 \cdot 10^{14}$. The conjecture that every odd number is the sum of three primes has been checked up to 10^{20}. It is equivalent to the statement that every integer greater than 17 is the sum of three *distinct* primes.

It is known that every integer is the sum of at most six primes, and that all sufficiently large numbers are the sum of a prime and a **semiprime**.

Also, every number n greater than 161 is the sum of distinct primes of the form $6n - 1$; if $n > 205$, of distinct primes, $6n + 1$; if $n > 55$, of distinct primes of the form $4n + 3$; and if $n > 121$, distinct primes, $4n + 1$.

If N is even and large enough, then there is a number n, less than N, such that $n(N - n)$ has at most five prime factors. (Halberstam and Richert 1974)

Many even numbers can simply be written as the sum of 3, 5, or 7 and another prime. The smallest that cannot is $98 = 19 + 79$.

Goldbach also conjectured in a letter to Euler of 1752 that every odd integer is the sum of a prime (which for Goldbach included the number 1) and double a square. Moritz Stern discovered in 1856 that 5777 and 5993 are exceptions.

Laurent Hodges has checked up to 1,000,000 and found no more exceptions, so there probably are none, for, as Euler pointed out, the number of representations in the form $N = p + 2a^2$ increases with N.

Hodges calculates, for example, that there are just twenty-eight numbers from 17 to 6797 that can be represented in only one way as $p + 2a^2$. (Hodges 1993)

Dan Zwillinger checked which even numbers from 2 to 1,000,000 can be written as the sum of **twin primes**, that is, primes that are each a member of a pair of twin primes. Only thirty-three exceptions were found, all less than 5000. The smallest three exceptions were 94, 96, and 98. The largest number than can be so represented in only one way is 24098. (Zwillinger 1979)

   See Hardy-Littlewood conjectures; Opperman's conjecture

good primes

Erdös and Strauss call a prime $p(n)$ good if $p(n)^2 > p(n-i)p(n+i)$ for all values of i from 1 to $n-1$.

   The infinite sequence of good primes starts, 5, 11, 17, 29, . . . (Guy 1994, 32)

Grimm's problem

Grimm conjectured that given any sequence of consecutive numbers, there is a matching sequence of *distinct* primes among their factors.

   For example:

| 1802 | 1803 | 1804 | 1805 | 1806 | 1807 | 1808 | 1809 | 1810 |

are divisible respectively by

| 53 | 601 | 41 | 19 | 43 | 139 | 113 | 67 | 181 |

It has been proved that there are only finitely many exceptions to the conjecture. (Grimm 1969)

Hardy, G. H. (1877–1947)

> The mathematician's patterns, like the painter's or the poet's, must be beautiful; the ideas, like the colors or the words must fit together in a harmonious way. Beauty is the first test: there is no permanent place in this world for ugly mathematics.
>
> —*G. H. Hardy*

Hardy was the greatest British mathematician of the twentieth century, and a brilliant exponent of very pure mathematics, especially prime numbers. In his book *A Mathematician's Apology* he put into words what so many mathematicians feel—"I am interested in mathematics only as a creative art"—and he expressed his satisfaction that nothing he had ever discovered or proved would have any practical use whatsoever: so he might have been disappointed to learn that the factorization of giant prime numbers is today important in **public key encryption**.

In an address to the British Association for the Advancement of Science he said, "A science is said to be useful if its development tends to accentuate the existing inequalities in the distribution of wealth, or more directly promotes the destruction of human life," adding that the study of prime numbers did neither, while "the greatest mathematicians of all ages have found in it a mysterious attraction impossible to resist." (Hardy 1915, 14)

In his inaugural lecture as professor at Oxford University, he said, "If I could attain every scientific ambition of my life, the frontiers of the Empire would not be advanced, not even a black man would be blown to pieces, no one's fortune would be made, and least of all my own. A pure mathematician must leave to happier colleagues the great task of alleviating the sufferings of humanity." (Hardy 1920)

Hardy once sent a postcard to a friend listing his six New Year wishes. One was to make 211 not out in the fourth innings of the last test match at the Oval. (Hardy was a cricket fanatic.) Another was to murder Mussolini, but the first was to prove the **Riemann hypothesis**.

Hardy's life was marked by two remarkable collaborations. The first started in 1911 when he began to work with J. E. Littlewood, who was eight years his junior. They worked together for thirty-five years writing nearly one hundred papers, many on number theory, including the Riemann zeta function.

Two years later, Hardy received a letter from **Ramanujan**, containing a large number of results, some well-known, some merely novel, and some, Hardy argued, so extraordinary that they could only be the work of a mathematician of genius.

He recorded this famous anecdote describing a visit to Ramanujan while he was ill:

> I had ridden in taxi cab number 1729 and remarked that the number seemed to me rather a dull one, and that I hoped it was not an unfavorable omen. "No," he replied, "it is a very interesting number; it is the smallest

number expressible as the sum of two cubes in two different ways." (Hardy 1940)

Ramanujan could have added that 1729 is also the third **Carmichael number**.

Hardy wrote two notable textbooks: *A Course of Pure Mathematics,* which introduced generations of English students to rigorous Continental ideas in analysis, and *An Introduction to the Theory of Numbers* (1938) written with E. M. Wright, which is now in its fifth edition and well worth study by any readers who want to go further into the technicalities of prime number theory.

Hardy-Littlewood conjectures

> We trust that it will not be supposed that we attach any exaggerated importance to the speculations which we have set out. . . . We have not forgotten that in pure mathematics, and in the Theory of Numbers in particular, "it is only proof that counts." It is quite possible, in the light of the history of the subject, that the whole of our speculations may be ill founded. Such evidence as there is points, for what it is worth, in the opposite direction.
>
> —*Hardy and Littlewood (1923)*

In one extremely long paper, Hardy and Littlewood published an influential series of **conjectures** about how numbers might be represented in different forms, none of which they proved, though they did give complicated formulae for the number of representations. Here are some of them.

- Conjecture B: There are infinitely many prime pairs n and $n + k$, for every even k.

- Conjecture E: The number of primes of the form $n^2 + 1$ is infinite and is equal to $cn^{1/2}/\log n$ as n tends to infinity, where c is approximately 1.3728134628 . . . (Finch 2003, 85)

  This remains only a conjecture, but it has been shown that there are an infinity of values of $n^2 + 1$ that are either primes or **semiprimes**.

  Hardy and Littlewood generalized this to "k-tuples" of the form $N, N + a_1, N + a_2, \ldots, N + a_k$, conjecturing that for any given $(a_1, \ldots, a_k)$ there are an infinite number of Ns for which the k-tuple $N, N + a_1, N + a_2, \ldots, N + a_k$ will be all prime, unless there is an elementary divisibility condition ensuring that one of these must be composite. The **twin primes** conjecture is the special case of N and $N + 2$.

- Conjecture H: Every large number is either a square or the sum of a prime and a square.
- Conjecture K: If k is any fixed number other than a (positive or negative) cube, then there are infinitely many primes of the form $m^3 + k$.
- Conjecture L: Every large number is either a cube or the sum of a prime and (positive) cube.
- Conjecture M: If k is any fixed number other than zero, there are infinitely many primes of the form $a^3 + b^3 + k$, where n and m are both positive.
- Conjecture N: There are infinitely many primes of the form $a^3 + b^3 + c^3$.
- Conjecture P: There are infinitely many prime pairs of the form $m^2 + 1$, $m^2 + 3$.

Despite the hesitation they expressed in the quotation above, these conjectures by two world-class mathematicians have stood the test of time—even as proofs are still lacking.

Hardy and Littlewood also conjectured that $p(x + y) \geq p(x) + p(y)$. This conjecture, however, is inconsistent with the conjecture that there are an infinite number of triples of primes of the forms $6n - 1$, $6n + 1$, $6n + 5$ and $6n + 1$, $6n + 5$, $6n + 7$. Which plausible conjecture is false? Probably the former. (Hardy and Littlewood 1923)

See formulae for primes; the prime numbers race

Primes of Polynomial Form

There is an infinite number of primes of the form $x^2 + y^4$ (Iwaniec and J. Friedlander) and of the form $x^3 + 2y^3$.

$n^4 + 1$ is prime for $n = 1, 2, 4, 6, 16, 20, 24, 28, 34, 46, 48, \ldots$ (Sloane A000068) (Lal 1967)

There is an infinity of primes of the form $n^2 + m^2$ or $n^2 + m^2 + 1$. If $b^2 - 4ac$ is a perfect square, then $an^2 + bn + c$ has algebraic factors. Otherwise, and if a, b, and c are coprime in pairs and if $a + b$ and c are not both even, then there is an infinity of primes $an^2 + bn + c$.

heuristic reasoning

> Now for the . . . heuristic argument. The reader should be warned
> that the following discussion involves some large leaps of faith.
>
> —*Robin Forman (1992)*

G. H. Hardy considered the number of ways in which an even number $n > 2$ can be written as the sum of two primes, and argued like this:

> If $m < n$ and n is large, the chance that m is prime is approximately 1: log n. If then we write n in every possible way in the form $m + m'$, the chance that both m and m' are prime is approximately 1: $(\log n)^2$. We should therefore expect the order of magnitude . . . to be $n/(\log n)^2$. (Hardy 1966)

This is an example of a *heuristic* argument. It is less than convincing because, as Hardy noted, the arithmetical form of n makes a difference to the answer. For a start, n must be even, unless n is 2 more than a prime.

It turns out that to get a more accurate result that takes into account the fact that n and $n + 2$ being prime are *not* actually independent events, we have to multiply this heuristic estimate by a factor 1.32032 . . . This makes a large difference in practice: the result is increased for large n by nearly one-third, yet the *heuristic* argument was certainly on the right lines and takes the mathematician a long way toward the correct conclusion.

(Brun proved that every even number is the sum of two numbers with at most nine odd prime factors, and that the number of such representations is also proportional to at least $n/(\log n)^2$.)

a heuristic argument by George Pólya

Pólya noted that while there are only eight pairs of twin primes less than 100, there are twice as many, sixteen, primes p such that $p + 6$ is also prime: 5-11, 7-13, 11-17, 13-19, 17-23, 23-29, 31-37, 37-43, 41-47, 47-53, 53-59, 61-67, 67-73, 73-79, 83-89, and 97-103. The proportion of 1:2 continues when higher primes are checked. Up to 30,000,000 there are 152,892 twin prime pairs and 304,867 pairs differing by 6.

Having noticed this fact, Pólya, writing in 1959 when computers were still quite weak, used data calculated by D. H. and Emma

Lehmer up to the limit $3 \cdot 10^7$ to compare the frequency of pairs, p and $p + d$, in which both numbers are prime.

He noticed that the frequency of pairs p and $p + d$ was roughly equal for $d = 2, 4, 8, 16, 32$, and 64. Similarly, the data for $d = 6, 12, 24$, and 48 were roughly equal, and so were those for $d = 10, 20$, and 40 and $d = 14, 28$, and 56.

Pólya continued with his observations, finally drawing the conclusion that if $\pi_d(x)$ is the number of pairs, p and $p + d$, both prime with $p < x$, then

$$\pi_d(x) \text{ is approximately equal to } \pi_2(x)\Pi(p - 1)(p - 2)$$

where the product is calculated over all the odd primes p that divide d.

Pólya then continued his argument by introducing an argument based on *probability*, and got the answer he wanted, *but only by an argument that he knew to be flawed*: to "correct" the flaw, he introduced an extra factor and got the "correct" result, while admitting that "I do not pretend to understand why the introduction of the upper bound . . . *should* yield the right result."

At this point we can leave George Pólya's brilliant speculations, while drawing the encouraging conclusion—which will astonish those people who falsely believe that mathematics is always rock-solid and purely logical—that such imaginative heuristic reasoning can lead to correct answers and so contribute to mathematical progress, by pointing toward more complete and final arguments. (Pólya 1959)

Hilbert's 23 problems

David Hilbert (1862–1943) was among the greatest mathematicians of all time. He is of most interest to us because of a famous speech that he delivered to the Second International Congress of Mathematicians in Paris on August 8, 1900, on "The Problems of Mathematics." With the confidence that characterized his own work—his cry, "We must know. We will know" is inscribed on his tomb in Göttingen—he proclaimed:

> This conviction of the solvability of any mathematical problem is a strong incentive in our work; it beckons us: this is the problem, find its solution. You can find it by pure thinking since in mathematics there is no Ignorabimus!

He then introduced his 23 famous problems. The eighth was *prime number problems*. Hilbert asked for a proof of the **Riemann hypothesis**, followed by an exact count of the number of prime numbers less than *N*, and an answer to the question whether "the occasional condensation of prime numbers which has been noticed in counting primes" is due to certain terms in one of Riemann's formulae.

Hilbert then turned to **Goldbach's conjecture** and the questions "whether there are an infinite number of pairs of prime numbers with the difference 2, or even the more general problem, whether the linear diophantine equation

$$ax + by + c = 0$$

(with given integral coefficients each prime to the others) is always solvable in prime numbers *x* and *y*."

The ninth problem was: proof of the most general reciprocity law in arbitrary number fields. This refers to generalizations of the law of **quadratic reciprocity**.

The tenth problem was: decision on the solvability of a Diophantine equation. This was eventually answered by **Matijasevic**, whose solution involved an extraordinary representation of the prime numbers.

See Goldbach's conjecture; Matijasevic; quadratic reciprocity, law of; Riemann hypothesis

home prime

Start with any number and form a new number by concatenating its prime factors, starting with the least. Then repeat: $30 = 2 \cdot 3 \cdot 5 \rightarrow 235 = 5 \cdot 47 \rightarrow 547$, which is prime, and the *home prime* of 30.

Similarly, the home prime of 4 is 211:

$$4 = 2 \cdot 2 \rightarrow 22 = 2 \cdot 11 \rightarrow 211$$

For $n = 2, 3, \ldots$ the home primes are 2, 3, 211, 5, 23, 7, 3331113965338635107, 311, 773, . . . (Sloane A037274)

Arguments from probability suggest that there is zero chance that any sequence will not contain a prime, but it has not been proved that no exceptional sequences exist.

The largest home prime, HP(n), for $n < 100$ is known to be HP(49) = HP(77), although it has not been calculated. The sequence starts, 49, 77, 711, 3379, 31109, 132393, 344131, . . . but calculation has stuck at step 100, which has a 204-digit composite cofactor whose prime factors are unknown.

The next largest HP(n) for $n < 100$ is HP(80) = 313,169,138, 727,147,145,210,044,974,146,858,220,729,781,791,489. (De Geest and Kruppa, *MathWorld*)

hypothesis H

Dickson's conjecture was generalized in 1958 by Sierpinski and Schinzel to polynomials in general. They called their generalization hypothesis H.

Suppose that you have a number of polynomials in x, each with no algebraic factors, with positive leading coefficients, *and such that there is no prime that necessarily divides their product, whatever the value of* x: then there is an infinity of values of x that makes them all prime simultaneously.

The italicized condition excludes cases such as $n^2 + 2$ and $n^2 + 1$, since their product, $(n^2 + 2)(n^2 + 1)$, will be even whether n is odd or even.

Hypothesis H implies there are infinitely many primes of the form $n^2 + 1$ *and* simultaneously of the forms $n^2 + 1$ and $2n^2 + 1$ (such as 37 and 73 when $n = 6$, or 577 and 1153 when $n = 24$).

It also predicts, for example, that there are infinity of integral Pythagorean triangles two of whose sides are primes, such as 3-4-5, 5-12-13, 11-60-61, and 19-180-181, since those sides will have the form $m^2 - n^2$, $2mn$, and $m^2 + n^2$, and the first and last expressions would be prime simultaneously. (Caldwell, *Prime Pages*)

illegal prime

Unauthorized playing of DVDs is illegal, so the Motion Picture Association of America has sued to stop the distribution of the DeCSS code, which can be used to unscramble the Content Scramble System used to protect DVDs. Not everyone is impressed, and this code, as well as others with the same function, has been published freely on the Internet and even printed on T-shirts. The code can also be transformed into a number. After all, computers work on binary bits, so that any program can be written as an integer.

Mathematics professor Phil Carmody has represented the DeCSS code as a prime, not because he doesn't believe in protecting artists' rights—he does—but because he thinks the current law favors multinational publishers against consumers. The result is a 1,401-digit prime that if transformed into hexadecimal would represent the DeCSS code. Courts in the USA have decided that this program and any version of it are an "illegal circumvention device" according to the Digital Millenium Copyright Act, or DMCA, because they allow copyright material to be shared without payment. Here it is:

485650789657397829309841894694286137707442087351357924019652073668698513401047237446968797439926117510973777701027447528049058831384037549709987909653955227011712157025974666993240226834596619606034851742497735846851885567457025712547499964821941846557100841190862597169479707991520048667099759235960613207259737979936188606316914473588300245336972781813914797955513399949394882899846917836100182597890103160196183503434489568705384520853804584241565482488933380474758711283395989685223254460840897111977127694120795862440547161321005006459820176961771809478113622002723448272249323259547234688002927776497906148129840428345720146348968547169082354737835661972186224969431622716663939055430241564732924855248991225739466548627140482117138124388217717602984125524464744505583462814488335631902725319590439283873764073916891257924055015620889787163375999107887084908159097548019285768451988596305323823490558092032999603234471140776019847163531161713078576084862236370283570104961259568184678596533310077017991614674472549272833486916000647585917462781212690073518309241530106302893295665843662000800476778967984382090797619859493646309380586336721469695975027968771205724996666980561453382074120315933770309949152746918356593762102220068126798273445760938020304479122774980917955938387121000588766689258448700470772552497060444652127130404321182610103591186476662963858495087448497373476861420880529443

Carmody has since transformed the banned CSS descrambler code into a 1,905-digit prime.

inconsummate number

It is a well-known rule that if the digit-sum of a number is 9, then 9 divides the number exactly, but this is exceptional. Most numbers, beyond the digits 1 to 9 themselves, are not divisible by their digit-sum. Those that are form this sequence:

n	10	12	18	20	21	24	27	30	36	40	42	45	48	50	...
n/(digit-sum)	10	4	2	10	7	4	3	10	4	10	7	5	4	10	...

Some numbers, m, are such that no number in base 10 is m times the sum of its digits. John Conway has labeled these the inconsummate numbers, and their sequence (Sloane A00365) for base 10 starts:

$$62, 63, 65, 75, 84, 95, 161, 173, 195, 216, 261, \ldots$$

They exist, of course, for other bases also. These are the smallest inconsummate numbers for bases 2 onward (Sloane A052491):

$$13, 17, 29, 16, 27, 30, 42, 46, 62, 68, 89, \ldots$$

induction

> [I]n [the higher] arithmetic the most elegant theorems frequently arise experimentally as the result of a more or less unexpected stroke of good fortune, while their proofs . . . elude all attempts and defeat the sharpest enquiries.
>
> —*Gauss (Laubenbacher and Pengelley 1994)*

Gauss in the extract above is describing one of the most delightful but also baffling features of prime numbers, which was especially obvious during the seventeenth and eighteenth centuries when basic and beautiful properties of the primes were lying around as it were like so many mathematical jewels waiting to be picked up by the keen-eyed mathematical naturalist.

Fermat and Euler and Gauss himself were brilliant calculators who noticed far more properties of the primes than they were able to prove, even after in some cases decades of effort. So Fermat spotted that the small **Fermat numbers** were prime, and then, as Gauss remarked, he "was misled by his induction and affirmed that all numbers contained in this form are necessarily prime." (Gauss 1801/1966, 459)

Gauss inferred when a teenager, from his statistical data, that $\pi(x)$, the number of primes less than or equal to x, is approximately $x/\log x$, but this was only proved by Hadamard and Vallée Poussin in 1896.

Very many results in the theory of numbers were originally claimed by the mathematician who discovered them, by experiment and observation or sometimes by "loose" but plausible reasoning, only to be proved (or disproved) later by someone else, which raises the question: how confident can you be—should you be—when you spot what appears to be a rock-solid pattern? Euler had one answer:

> For each of us can convince himself of this truth by performing the multiplication as far as he may wish; and it seems impossible that the law which has been discovered to hold for 20 terms, for example, would not be observed in the terms that follow. (Euler: Wells 1997, 64)

However, we can pick two possible holes in that reasoning. Euler was talking about a series and in series or sequences such as

$$1, 3, 6, 10, 15, 21, 28, \ldots$$

the pattern does seem to be extremely strong—this seems obviously to be the triangular numbers—and yet, consulting that invaluable aid to the modern mathematical naturalist, Neil Sloane's *On-Line Encyclopedia of Integer Sequences*, there is a good chance that we shall find more than one sequence that fits these seven starting values. So let's go to www.research.att.com/~njas/sequences, select the "sequence" option, and type in, 1, 3, 6, 10, 15, 21, 28.

What do we get? The first result is, naturally, sequence A000217, the triangular numbers, with their formula, $n(n + 1)/2$, and a list of many ways in which they arise. They are, 0, 1, 3, 6, 10, 15, 21, 28, 36, 45, 55, 66, 78, 91, 105, 120, . . .

However, this entry is followed by many others that are not the triangular numbers, but look much like them. Here are some of them.

- The sequence 0, 0, 1, 3, 6, 10, 15, 21, 28, 35, 43, 52, 62, 73, 85, 98, which is "number of edges in 8-partite Turán graph of order n." This is followed by similar results for 9-, 10-, 11-, and 12-partite graphs.

- Next is sequence A037123: 1, 3, 6, 10, 15, 21, 28, 36, 45, 46, 48, 51, 55, 60, 66, 73, 81, 90, 100, which has the recurrence relationship $a(n) = a(n - 1)$ + sum of digits of n.
- Sequence A061791: 1, 3, 6, 10, 15, 21, 28, 36, 45, 55, 66, 77, 90, in contrast, is the number of distinct sums $i^3 + j^3$ for $1 \leq i \leq j \leq n$.
- Sequence A061076: 1, 3, 6, 10, 15, 21, 28, 36, 45, 45, 46, 48, 51, 55, 60, 66, 73, 81, 90, 90, 92, 96, which is the sum of the products of the digits of all the numbers from 1 to n. (Murthy: Conroy)

There are other sequences that start 1, 3, 6, 10, 15, 21, 28. So we have to fall back on Euler's reference to "20 terms," as if that were far enough to provide complete confidence and total satisfaction. Major Percy MacMahon remarked of one of his conjectures, "This most remarkable theorem has been verified as far as the coefficient of x^{89} by actual expansion so that there is practically no reason to doubt its truth; but it has not yet been established." (Pólya 1954, 96)

Wise words: the belief that $2^{n-1} - 1$ is only divisible by n if n is prime is true up to $n = 340$, but then $n = 341$ is the first exception, and the first (Fermat) **pseudoprime**.

D. H. and Emma Lehmer had to go much further to discover that $2^n \equiv 3 \pmod{n}$ is first satisfied for $n = 4,700,063,497$. In other words, if you continue this sequence of the remainders when $2^n - 3$ is divided by n, from $n = 2$ onwards,

$$1 \quad 2 \quad 1 \quad 4 \quad 1 \quad 6 \quad 5 \quad 5 \quad 1 \quad 10 \quad \ldots$$

then you only have 4.7 billion and a bit terms to go before the first zero appears.

Another misleading result is that $\phi(30n + 1) > \phi(30n)$ for all values of n up to 20,000,000, yet it has been proved that eventually there is an n for which $\phi(30n) > \phi(30n + 1)$. (Newman 1997, 256–57)

G. H. Hardy remarked of **Goldbach's conjecture**, "The numerical evidence for the truth of the [conjecture] is overwhelming. It has been verified up to 1000 by Cantor and . . . up to 10,000 by Haussner." Today, Goldbach's conjecture has been checked up to $4 \cdot 10^{14}$, but as mathematicians know only too well, infinity is much bigger than 10^n.

See conjectures; errors; Riemann hypothesis; strong law of small numbers

jumping champion

An integer is a jumping champion if it is the most frequently occurring difference between consecutive pairs of primes. The idea was suggested by Harry Nelson in 1978–1979, though the name was given these numbers by John Conway in 1993. (Nelson 1978–79)

There are occasionally several jumping champions in a range. The champion is 2 for these ranges of numbers: 7–100, 103–106, 109–112, . . . Both 2 and 4 are joint champions for the ranges, 101–102, 107–108, 113–130, . . .

For $n \leq 1000$, the champion is mostly 2, 4, or 6. The champion up to about 1.74×10^{35} is 6, after which 30 predominates. The next champion, 210, enters at about 10^{425}. Since 2, 6, 30, 210, . . . is the sequence of **primorials**, it is naturally conjectured that the sequence of jumping champions consists of 1 and 4, and the primorials, the next being $11\# = 2310$. (Odlyzko, *MathWorld*)

k-tuples conjecture, prime

How many triples of consecutive primes are there? We exclude 3-5-7 because it is one of a kind. Also, the remainders when we divide the pattern n, $n + 2$, and $n + 4$ by 3 must be 0, 2, and 1 in that cyclic order, so one of them is always divisible by 3.

So we focus on triples such as 5-7-11 and 7-11-13, of the form n, $n + 2$ *or* $n + 4$, $n + 6$. They exist for these values of n:

5, 7, 11, 13, 17, 37, 41, 67, 97, 101, 103, 107, 191, 193, 223, 227, 277, 307, 347, 457, 613, 641, . . . (Sloane A007529)

Of these, ten are of the n, $n + 2$, $n + 6$ form and twelve are of the n, $n + 4$, $n + 6$ form. Up to 10^8 there are 55,600 triples of the form $(p, p + 2, p + 6)$ and 55,556 of the form $(p, p + 4, p + 6)$. (Caldwell, *Prime Pages*) The Hardy-Littlewood prime *k*-tuples conjecture implies that the total is about 55,490.

In a different kind of triple, the middle prime is the average of its nearest neighbors. These are the middle primes:

5, 53, 157, 173, 211, 257, 263, 373, 563, 593, 607, 653, 733, . . . (Sloane A006562)

It is not known if there is an infinity of quadruples of primes, or of *k*-tuples in general, in arithmetic progression. Prime quadruplets, sometimes known as a *prime decade* because their final digits in base 10 must always be 1, 3, 7, and 9, cannot have the form *n*, *n* + 2, *n* + 4, *n* + 6 for the same reason as before—one of these numbers must be a multiple of 4.

We also exclude 2-3-5-7 and 3-5-7-11, which both include the "pathological" triple, 3-5-7, not to mention the unique even prime!

So we consider the set, *n*, *n* + 2, *n* + 6, and *n* + 8, and the quadruplets that start with 5-7-11-13, and continue with 11-13-17-19; 101-103-107-109; 191-193-197-199; 821-823-827-829.

Subsequent sets start with these values of *n*: . . . 1481, 1871, 2081, 3251, . . . (Sloane A007530) The sequence includes the smallest prime quadruplet of fifty digits. (Stevens 1995, 17–22)

10000000000000000000000000000000000000000058537891,
10000000000000000000000000000000000000000058537893,
10000000000000000000000000000000000000000058537897,
10000000000000000000000000000000000000000058537899

Tony Forbes keeps details of the largest known prime *k*-tuples at his site, www.ltkz.demon.co.uk/ktuplets.

See Dickson's conjecture; Hardy-Littlewood conjectures

knots, prime and composite

A prime number is one that cannot be "composed" of two other numbers, so it is no surprise that "prime" is used metaphorically to describe other mathematical objects that cannot be fitted together from other (prime) pieces.

Composite knots can be formed by joining together on the same string two other knots, called *factor* knots. Knots that are not composite are *prime*. There is even a prime knots factorization theorem. The ordinary trefoil knot is prime:

These are the seven prime alternating knots with seven crossings:

This figure shows how you can "add" two prime knots to create a composite knot:

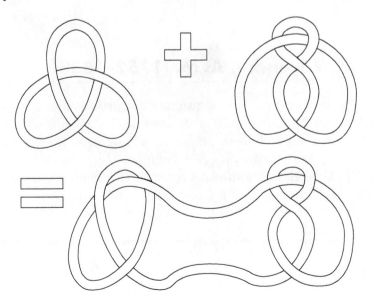

To factorize the right-hand knot, cut it at the narrow "neck" and join the loose ends to re-create the two prime knots on the left.

Landau, Edmund (1877–1938)

In 1903 Landau published a proof of the **prime number theorem** that was simpler than those of Hadamard and de la Vallée Poussin.

Then in an address at the Fifth International Congress of Mathematicians in Cambridge, England, in 1912 he described four problems that he said were "unattackable" in the present state of knowledge. These were **Goldbach's conjecture**, the **twin primes** conjecture, the existence of an infinity of primes of the form $n^2 + 1$, and the problem of the existence of a prime between n^2 and $(n + 1)^2$.

left-truncatable prime

Left-truncatable primes remain prime when the left-hand digit is repeatedly removed. According to Chris Caldwell there are 4,260 such primes, of which the three longest are

> 959 18918 99765 33196 93967 . . . (23 digits)
> 966 86312 64621 65676 29137 . . . (23 digits)
> 3576 86312 64621 65676 29137 . . . (24 digits)

(Caldwell, *Prime Pages*)

Legendre, A. M. (1752–1833)

Legendre discussed the law of **quadratic reciprocity**, and mistakenly believed that he had proved a theorem that was later proved by Dirichlet: in every arithmetic sequence whose terms do not have a common factor, there are an infinite number of primes.

In 1823 Legendre published his proof that Fermat's equation, $x^5 + y^5 = z^5$, has no solution in integers, and he proved that every number not of the form $8n + 7$ is the sum of three squares.

He also produced one of the earliest estimates of $\pi(x)$, conjecturing in 1798, and again in 1808, that $\pi(x)$, the number of primes less than x, tends to

$$\pi(x) = x/(\log(x) - A)$$

as x tends to infinity: he supposed from his experimental evidence that the best value of A was 1.08366 ... known as Legendre's constant, which he published in his *Théorie des Nombres* in 1798. In 1850, Tchebycheff showed that the best figure, if the limit exists, is 1, but he was unable to prove that the limit did indeed exist.

Lehmer, Derrick Norman (1867–1938)

D. N. Lehmer was an unusual mathematician: he published and edited poetry and studied North American Indians, collecting their legends and stories, as well as their music, which he used to compose three musical works based on Indian culture.

In 1914 he published a *List of prime numbers from 1 to 10,006,721,* unusually listing 1 as a prime. He had already published in 1909 a table in book form of the largest factors of all the numbers up to 10,017,000, excluding numbers with the obvious factors, 2, 3, 5, and the less obvious 7. He was also a founder of the **Cunningham project**.

Lehmer, Derrick Henry (1905–1991)

Derrick Henry Lehmer continued his father's work on prime numbers, aided by his wife and coworker Emma Lehmer (1906–), who was originally his father's graduate assistant. At age ninety, Emma Lehmer has finished her research work, written up her husband's unfinished work, and overseen the publication of these materials.

D. H. Lehmer was one of the founders, in 1943, of the journal *Mathematical Tables and other Aids to Computation*, which in 1959 became *Mathematics of Computation*. In his PhD thesis he proposed the **Lucas-Lehmer test** for Mersenne primes. He also worked on ENIAC, the first electronic computer in the United States. He was fascinated from his student days with number sieves, electromechanical machines that tested numbers to see if they satisfied certain congruences.

Lehmer's machines searched a sequence of numbers, checking automatically, by mechanical, electrical, or optical means or a combination of methods, for numbers that left the required remainder when divided by certain primes.

Hunting Big Game in the Theory of Numbers

by Derrick N. Lehmer

On the 19th of October a little group of mathematicians gathered in the Burt Laboratories in Pasadena, California, around a mysterious machine to watch it attack a problem in mathematics. It was a simple enough problem to state. It had only to find two numbers which when multiplied together would give 5,283,065,753,709,209. Any person with a few hundred years of leisure time on his hands could work it out. . . .

And after all we had taken only an important outwork in the assault upon a real fortress. This victory had merely cleared the decks for action against another and much larger number which was under grave suspicion of being a prime; that is, not the product of any two smaller numbers. This number is the great unconquered factor of $2^{95} + 1$. It is the nineteen digit number 3,011,347,479,614, 249,131.

On the 9th of October, just after the discovery of the "imp" the machine was set to do some real work in the theory of numbers. There was a large factor of $2^{93} + 1$, namely the number 1,537,228,672,093,301,419 which was known by a very powerful test to be composite, but the test would not furnish the factors. The smallest factor was known also to be larger than 300,000, and might be large enough to occupy the time of a skilled computer for over twenty-five years to find it. (*Scripta Mathematica*, September 1932)

His first electromechanical number sieve was constructed in 1926, using nineteen rotating bicycle chains. In 1932 he built a larger machine using thirty-two gear wheels, which could check 300,000 numbers every minute. These devices were followed by others using 16mm film, then vacuum tubes and delay lines.

The 1932 geared sieve was demonstrated at the Chicago Century-of-Progress Expositions. Sieves are still in use, for example at the University of Manitoba, where Fung and Williams used one to find quadratics with many prime values (*see* **Euler's quadratic**). Their sieve solves 133,000,000 linear congruences per second. (Rubinstein n.d.) (Fung and Williams 1990, 346)

A Lehmer Conference was held at the end of August 2000 at the

University of California at Berkeley to honor the work and influence of all three members of the Lehmer family.

See Cunningham project; Lehmer, Derrick Norman

Linnik's constant

Dirichlet's theorem proved that there are an infinite number of primes in the arithmetic sequence,

$$a, a + b, a + 2b, a + 3b, \ldots, a + nb$$

but when does the first prime occur? Yuri Vladimirovich Linnik (1915–1972) showed in 1944 that there is a number L, Linnik's constant, such that the first prime is at most a^L, independently of the value of b, provided it is large enough.

Roger Heath-Brown proved in 1992 that $L \leq 5.5$, but it is conjectured that $L = 2$, which is certainly the correct figure for almost all values of b.

Vinogradov's celebrated proof of the ternary **Goldbach's conjecture** states that all large enough odd numbers can be written as the sum of three primes. Linnik showed that for large n we can write n as the sum of two primes and k powers of 2, where k is now known to be at most 12.

Liouville, Joseph (1809–1882)

Liouville, among many achievements, proved the following extraordinary property of the divisors of the divisors of a number:

Start with a number, say, 10, and write down its divisors: 1, 2, 5, 10. Now write down the number of divisors of each divisor. They have 1, 2, 2, and 4 divisors, respectively.

Then: $(1 + 2 + 2 + 4)^2 = 81 = 1^3 + 2^3 + 2^3 + 4^3$

Similarly, the divisors of 24 are 1, 2, 3, 4, 6, 8, 12, and 24, and the number of their divisors are 1, 2, 2, 3, 4, 4, 6, and 8, respectively, and

$$(1 + 2 + 2 + 3 + 4 + 4 + 6 + 8)^2$$
$$= 1^3 + 2^3 + 2^3 + 3^3 + 4^3 + 4^3 + 6^3 + 8^3 = 900$$

This theorem also implies that if N is double an odd number and has no factors of the form $4n + 3$ (that remarkable distinction again!) and if we write down the ways in which its even factors can be expressed as the sum of two odd squares, then the sum of the cubes of the numbers of ways will equal the square of their sum. For example,

$$50 = 1^2 + 7^2 = 7^2 + 1^2 = 5^2 + 5^2$$
$$10 = 3^2 + 1^2 = 1^2 + 3^2 \text{ and } 2 = 1^2 + 1^2$$

and so: $\quad 3^3 + 2^3 + 1^3 = (3 + 2 + 1)^2 = 6^2$

(Dickson 1952, vol. 1, 286)

Littlewood's theorem

John Edensor Littlewood (1885–1977) is best known for his long collaboration with G. H. **Hardy** on analytic number theory.

In 1914 he proved a very surprising theorem about two functions for the number of primes less than or equal to x, $\pi(x)$, and $\text{Li}(x)$. According to the table on p. 185, $\pi(x)$ is always less than $\text{Li}(x)$, and indeed it is known that $\pi(x) < \text{Li}(x)$ for all $x \leq 10^{12}$. Nevertheless Littlewood showed that $\pi(x) < \text{Li}(x)$ is false infinitely often, despite the apparently overwhelming numerical evidence to the contrary.

It is now known that there is a counterexample below 10^{400}.

This proof that a function which is related to the **Riemann hypothesis** can switch in value so far from the origin naturally undermines the evidence that Riemann's hypothesis is true, based on the calculation of its first few billion zeros. Who knows what might happen a few billion farther on?

the prime numbers race

A similar phenomenon appears in the prime numbers race, though the switch occurs much sooner. All primes from 5 onward are of one of the forms $4n + 1$ or $4n + 3$. If we start counting the number in each class, it seems that $4n + 3$ always wins, but this is misleading: at 26861 the lead switches. Up to that limit, there are 1,473 primes of the form $4n + 1$ and only 1,472 of the form $4n + 3$. (Leech 1957)

The succession of primes $4n + 1$ and $4n + 3$ is very irregular, as might be expected. Thus the race between primes $4n + 1$ and $4n + 3$ is tied at: 2, 5, 17, 41, 461, 26833, 26849, 26863, 26881, 26893, 26921,

615769, 616793, 616829, 616843, . . . (Sloane A007351) The prime 26861 is of the form $4n + 1$ and puts $4n + 1$ into the lead for the first time, but the lead is immediately lost.

The repeated equality at five values between 26833 and 26921, followed by a large gap, suggests some resemblance between this prime race and a random walk. As long as the random walk is near the origin, it has a significant chance of returning to the origin in the near future. The farther it moves away, the longer it is likely to stay away.

For most of the first few billion numbers, $4n + 3$ is winning over $4n + 1$. The sixth and largest known region for which $4n + 1$ is leading stretches from 18,465,126,293 to 19,033,524,538.

The sequence

$$11593 \quad 11597 \quad 11617 \quad 11621 \quad 11633$$
$$11657 \quad 11677 \quad 11681 \quad 11689$$

is a sequence of nine consecutive primes of the form $4n + 1$, discovered by Den Haan, and 241603 is the start of thirteen consecutive primes of the form $4n + 3$. (Guy 1994, 13)

The race between $3n + 2$ and $3n + 1$ initially favors $3n + 2$, though there is an infinity of values of n for which $3n + 1$ is leading, starting with 608,981,813,029, found by Carter Bays and Richard Hudson on Christmas Day 1976. The lead then changes repeatedly just above $608,981,813 \cdot 10^3$. It can be proved that the lead in all such prime number races changes infinitely often.

See induction

Lucas, Édouard (1842–1891)

Lucas worked at the Paris Observatory as assistant to Le Verrier (who had predicted the position of the new planet Neptune) and fought in the Franco-Prussian War (1870–71) in the artillery. He is reputed to have lived in Paris in a house on the site of the house where Pascal died. He himself died a bizarre death. At a banquet, a plate was dropped and a chip flew up and cut his cheek. He died soon after from erysipelas at the age of forty-nine.

He was deeply interested in number theory and its history. He was a member of the committee that edited Fermat's works for publication, and he wrote papers on astronomy, geometry, analysis, combinatorics, and calculating devices, as well as number theory. (Williams 1998, 53)

He also wrote several papers on the theory of weaving, a subject that has a surprising connection with computing: the problem of producing the most complex patterns in woven silk was solved by Jacquard by using punched cards that selected the threads used during the weaving of a particular pattern, making the Jacquard loom the first "stored program" machine. Charles Babbage also used punched cards to instruct his Difference Engine, which he used to calculate the values of **Euler's quadratic**, and these in turn were the ancestors of the Holerith cards first used in a United States Census and later in electronic computers. (Williams and Shallit 1994)

In 1879 Lucas started a column on recreational mathematics in *Revue Scientifique de la France et de L'Etranger*, and later in *La Nature*, and between 1882 to 1894 published a four-volume *Récréations Mathématiques*, which included **magic squares** and the Tower of Hanoi, a puzzle that he invented in 1883 and published under the anagram "N. Claus de Siam" (or Lucas d'Amiens) as "a game of combination designed to explain the binary system of numeration," as the inside cover of the version sold to the public explained.

The object is to transfer all the rings from peg A to peg C, following the rules that:

1. Only one ring may be moved at a time.
2. No ring may be placed on top of a smaller ring.

The puzzle created a sensation and was soon exploited in an advertising version, the Eight Puzzle, in which each disk advertised a different product. Lucas also made a giant version more than a meter high for public display. It was sold with a story of the priests at the

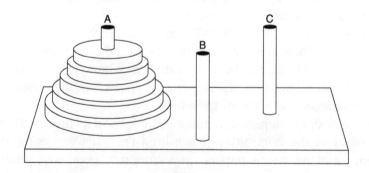

Temple of Benares, where there is, "beneath the dome which marks the center of the world, a brass plate in which are fixed three diamond needles. On one of these needles God placed at the Creation sixty-four disks of pure gold. . . . Day and night unceasingly the priests transfer the disks from one diamond needle to another." When the sixty-four disks have all been transferred, relates Lucas, "then towers, temple, and priests alike will crumble into dust, and with a thunderclap the world will vanish."

The point of his fable is that the minimum number of moves needed to transfer n disks from one peg to another specified peg is $2^n - 1$, a **Mersenne number**. (The method of moving the disks to minimize the moves required can also be represented by using binary numbers.)

Lucas noted in the rules that came with the puzzle that "at one move per second, it takes more than four minutes to move a tower of eight rings," which requires 255 moves. Contemporary illustrations of the puzzle show ten rings, or a version with five pegs and sixteen rings. The version using only three pegs would take more than eighteen hours, supposing that no mistakes were made on the way! To transfer sixty-four disks would take $2^{64} - 1$ moves. At one move per second, the priests would take more than 500 billion years.

All the possible moves in the Tower of Hanoi puzzle can be represented in this diagram, which is for three disks only, and resembles the pattern of **Pascal's triangle** and Sierpinski's gasket.

The notation "231," for example, means that the smallest disk is on peg 2, the middle disk on peg 3 and the largest disk is on peg 1. If

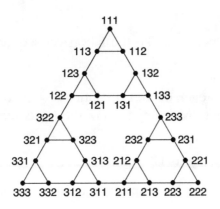

all three disks start on peg 1, then they can be moved to peg 2 in seven moves by following the sequence down the right side of the triangle.

The figure will always split up into three smaller triangles (and then into smaller triangles still) because the shortest way to transfer n disks from peg 1 to peg 2 is always to move $n - 1$ disks from peg 1 to peg 3, then move the largest disk to peg 2, and then move $n - 1$ disks from peg 3 to peg 2.

the Lucas sequence

Lucas also wrote a book on the *Theory of Numbers* and studied the Fibonacci sequence and the related sequence that is now called the Lucas sequence:

n	1	2	3	4	5	6	7	8	9	10	11 ...
$L(n)$	1	3	4	7	11	18	29	47	76	123	199 ...

(It sometimes is started: 2, 1, 3, 4, 7, 11, . . .)

It has the same rule of formation: each term is the sum of the previous two terms, so it is an example of a generalized Fibonacci sequence.

If the first two terms are labeled L_0 and L_1, Lucas proved that if p is prime, then $L_p \equiv 1 \pmod{p}$. This is not a test for primality, however, because the converse is sometimes false: $L_{705} \equiv 1 \pmod{705}$ although 705 is composite. This conclusion is reminiscent of **Fermat's Little Theorem** and so the exceptions to the converse to Lucas's rule are called Lucas pseudoprimes. The only ones less than 100,000 are 705, 2465, 2737, 3745, 4181, 5777, and 6721. (Singmaster 1983)

The Lucas sequence has, like the Fibonacci sequence, many other properties:

Whereas every integer divides an infinity of Fibonacci terms, the number 5 does not divide any term of the Lucas sequence:

$$L_{2n} = L_n^2 - 2(-1)^n$$
$$L_n^2 = L_{n-1}L_{n+1} + 5(-1)^n \quad \text{and} \quad L_nL_{n+3} = L_{n+1}L_{n+2} + 5(-1)^n$$

. . . illustrating that 5 has a special role in the Lucas sequence also.

The Fibonacci and Lucas sequences are also related: for example,

$$L_n = F_{n-1} + F_{n+1} \quad \text{and} \quad F_{2n} = F_n L_n$$
$$L_{m+n} = \tfrac{1}{2}(5F_m F_n + L_m L_n) \quad \text{and so} \quad L_{2n} = \tfrac{1}{2}(5F_n^2 + L_n^2)$$

L_n divides L_m if and only if n divides m and m/n is odd.

$2 \mid L_n$ if and only if $3 \mid n$; $7 \mid L_n$ if and only if n is an odd multiple of 4; $11 \mid L_n$ if and only if n is odd multiple of 5.

The Lucas sequence has periodic final digits with period 12: they go

$$1 \quad 3 \quad 4 \quad 7 \quad 1 \quad 8 \quad 9 \quad 7 \quad 6 \quad 3 \quad 9 \quad 2$$

The final two digits are periodic with period 60.

Like the Fibonacci sequence, the ratio L_{n+1}/L_n tends to the Golden Ratio, $\phi = \tfrac{1}{2}(\sqrt{5} + 1) = 1.618 \ldots$ Moreover, L_n is the closest integer to ϕ^n. For example, $\phi^9 = 76.01315 \ldots$ and $L_9 = 76$.

The sequence of prime Lucas numbers starts: 3, 7, 11, 29, 47, 199, 521, 2207, 3571, 9349, and then the very large 3010349, 54018521, 370248451, 6643838879, . . . (Sloane A005479)

Lucas also considered more general sequences that he used in his work on primality testing. Binet had proved in 1843 that the Fibonacci numbers F_n can be defined by this equation:

$$F_n = \frac{a^n - b^n}{a - b}$$

where a and b are $(1 + \sqrt{5})/2$ and $(1 - \sqrt{5})/2$, the roots of the equation $x^2 - x - 1 = 0$. Although these roots are irrational, these functions of them are always integers—the numbers in the Fibonacci sequence.

Lucas considered instead *any* equation $x^2 - Px + Q = 0$, where P and Q are coprime. If its roots are a and b, then he defined

$$U_n = \frac{(a^n - b^n)}{a - b}$$

and also
$$V_n = a^n + b^n$$

Once again, even if a and b are irrational, U_n and V_n will always be integers, and they satisfy the same recurrence relationship as the Fibonacci numbers: each term is the sum of the previous two terms.

What is the point of these definitions? What are these *generalized Fibonacci numbers* useful for? It was the genius of Lucas to realize, and to show in a long paper published in 1878, that they were related to a wealth of problems, including, as he wrote, "the theory of determinants, combinations, continued fractions, divisibility, divisors of quadratic forms . . . quadratic residues, decomposition of large numbers into their prime factors, etc." (Williams 1998, 71) He also pointed out that they behaved in some ways like the trigonometrical functions, sine and cosine! They could also be used for **primality testing**.

primality testing

Lucas was the first mathematician to realize and demonstrate that even very large numbers could be tested for primality without recourse to tedious and impracticable trial division. He did this in a series of thirteen papers all written between 1875 and 1878. (During the same period he wrote more than seventy papers on various other subjects—extraordinary productivity.)

By 1877, he had factorized the first sixty **Fibonacci numbers**. (Mollin 2001, 27–28) In 1876 he showed that the gigantic (for that era) **Mersenne number** $M_{127} = 2^{127} - 1$ is prime. He used this test, which required a great deal of calculation:

Let $p = 2^{4m+3} - 1$, where $4m + 3$ is prime. Form the sequence,

$$3, 7, 47, 2207, 4870847, \ldots$$

in which $S_1 = 3$ and $S_{n+1} = S_n^2 - 2$. Then p is prime if the least value of k such that $p \mid S_k$ is $4m + 2$.

This test is simple in theory but requires an enormous amount of computation. How did Lucas do it, by hand? With typical ingenuity, he used binary arithmetic and turned the calculation into a kind of game, exploiting the fact that multiplication is mod p, using a 127-by-127 chessboard. This is how he explained his method, adapted from the account by Williams and Shallit (1994, 491–92), using the simpler example of M_7.

Lucas's game of calculation

Consider $M_7 = 2^7 - 1 = 127$. We have $S_1 = 3$, $S_2 = 7$, $S_3 = 47$, $S_4 = 47^2 - 2 \equiv 48 \pmod{127}$, $S_5 = 48^2 - 2 \equiv 16 \pmod{127}$, $S_6 = 16^2 - 2 \equiv 0 \pmod{127}$.

So $127 \mid S_6$ but no smaller term, and so 127 is prime.

Next, Lucas noted that if $M_n = 2^n - 1$, then $2^{n+m} \equiv 2^m \pmod{M_n}$. In this case, $M_n = 127 = 2^7 - 1$ and so $2^{7+m} \equiv 2^m \pmod{127}$.

The main operation of testing involves squaring, subtracting 2, and reduction modulo 127. To perform this on S_3 to calculate S_4, we first write S_3 in binary, as 101111. Next, we start to square it by long multiplication:

$$
\begin{array}{r}
1\ 0\ 1\ 1\ 1\ 1 \\
1\ 0\ 1\ 1\ 1\ 1 \\
\hline
1\ 0\ 1\ 1\ 1\ 1 \\
1\ 0\ 1\ 1\ 1\ 1 \\
1\ 0\ 1\ 1\ 1\ 1 \\
1\ 0\ 1\ 1\ 1\ 1 \\
0\ 0\ 0\ 0\ 0\ 0 \\
1\ 0\ 1\ 1\ 1\ 1 \\
\hline
\end{array}
$$

However, because we only need to find the answer modulo 127, and we know that $2^{7+m} \equiv 2^m \pmod{127}$, so $2^7 \equiv 2^0$, $2^8 \equiv 2^1$, $2^9 \equiv 2^2$, and so on, we can put the six lines of the long multiplication into this square array:

No. of column		7 6 5 4 3 2 1
		0 1 0 1 1 1 1
		1 0 1 1 1 1 0
		0 1 1 1 1 0 1
		1 1 1 1 0 1 0
		0 0 0 0 0 0 0
		1 1 0 1 0 1 1

Lucas suggested using part of a chessboard, with pawns for ones and empty squares for zeros. Now that the pieces are arrayed, we follow these two rules:

1. Take (when possible but only once) a pawn away from column 2. This corresponds to subtracting 2 from the square. If a pawn never appears in column 2, then 2 must be subtracted from the final answer.

2. For each pair of pawns in any column, remove one from the board and move the other into the column to the left, remembering that the column to the left of column 7 is column 1.

Continue performing these operations until the only pawns remaining are in the first row. In our case we get the first row to be,

No. of column   7 6 5 4 3 2 1
                       0 1 1 0 0 0 0

which represents $S_4 \equiv 48$ in decimal.

Lucas claimed that these operations, with practice, could be performed very quickly, and this was how he showed that M_{127} is prime. On the other hand, it would seem easy to make a small mistake, with no record left behind to be checked, so this no doubt explains why Lucas always showed a slight uncertainty as to whether he really had proved M_{127} prime.

the Lucas-Lehmer test

The converse of **Fermat's Little Theorem** is false: there are numbers that satisfy it but are not prime. However, Lucas in 1876 added an extra condition to create a test for primality. Suppose we want to know if n is prime. If there is a number a such that $a^{n-1} \equiv 1$ (mod n) (so n satisfies Fermat's Little Theorem), and also a^m is *not* congruent to 1 (mod n) for $m = 1$ to $n - 2$, then n is prime.

This took a lot of calculation, however, especially before modern computers were invented, so in 1891 Lucas shortened it. It was then improved by Lehmer and Kraitchik. This is the result:

Suppose than $n > 1$. Then if for every prime factor q of $n - 1$ there is an integer a such that $a^{n-1} \equiv 1$ (mod n), and $a^{(n-1)/q}$ is not congruent to 1 (mod n), then n is prime.

This test depends on the complete factorization of $n - 1$. **Pockington's theorem** is an improvement that requires only the partial factorization of $n - 1$.

See Euler's quadratic; Fibonacci numbers; Lucas-Lehmer test; primality testing

lucky numbers

In 1955 Stanislav Ulam (1909–1984) and three colleagues at the Los Alamos Scientific Laboratory used an early computer to investigate a sequence that strongly resembles the prime numbers.

The sieve of **Eratosthenes** leaves behind the prime numbers. It can easily be modifed to leave behind different sequences of numbers. Write down the natural numbers and delete every second number:

$$1 \quad 3 \quad 5 \quad 7 \quad 9 \quad 11 \quad 13 \quad 15 \quad 17 \quad 19 \quad \ldots$$

The first number remaining is 3, so delete every third number, starting with 5. This leaves:

$$1 \quad 3 \quad 7 \quad 9 \quad 13 \quad 15 \quad 19 \quad 21 \quad 25 \quad 27 \quad 31 \quad 33 \quad 37 \quad 39 \quad \ldots$$

(This means that all numbers of the form $3n + 2$ have been struck out.) The first "unused" number is now 7, so delete every seventh number. The next "unused" number is 9, so delete every ninth number. And so on. The final result is the sequence of *lucky numbers*:

1, 3, 7, 9, 13, 15, 21, 25, 31, 33, 37, 43, 49, 51, 63, 67, 69, 73, 75, 79, 87, 93, 99, 105, 111, 115, 127, 129, 133, 135, 141, 151, 159, 163, 169, 171, 189, 193, 195, 201, 205, 211, . . . (Sloane A000959)

The largest known lucky number is 9,999,999,997. (Schneider 2002) There are "fewer" lucky numbers than prime numbers, meaning that although there is an infinity of each, the prime numbers are denser; they occur in any largish finite interval with greater frequency.

Ulam and his colleagues used an "electronic computing machine" to find the lucky numbers up to 48,000. There were 4,523, compared to 4,947 primes in the same interval. They noted that the numbers of luckies of the forms $4n + 1$ and $4n + 3$ were roughly equal, as they are for the primes. Likewise, the gaps between successive lucky numbers seem to roughly match those between primes, and every even integer between 1 and 100,000 is the sum of two lucky numbers. The number of lucky twin pairs (differing by 2) is roughly that of the twin primes. They also noticed that there are 715 numbers that are both prime and lucky, up to 48,600 (though no one knows if there is an infinity of prime lucky numbers). (Gardiner, Lazarus, Metropolis, and Ulam 1956)

Walter Schneider in May 2002 calculated all lucky numbers up to 10^{10} and verified **Goldbach's conjecture** to the same limit.

the number of lucky numbers and primes

limit	luckies	primes	twin luckies	twin primes
10^3	153	168	33	35
10^4	1,118	1,229	178	205
10^5	8,772	9,592	1,162	1,224
10^6	71,918	78,498	7,669	8,169

(Schneider 2002)

"random" primes

These parallels between the lucky numbers and the primes suggest that many of the properties of the prime numbers depend only on the fact that they can be created by a sieve. This conclusion is reinforced by other methods of creating "prime" numbers. For example, Harald Cramér in 1936 imagined putting black and white balls into a sequence of jars representing the numbers 1, 2, 3, 4, . . . so that the probability of drawing a white ball from the kth jar will be $1/\log k$. Then he chose one ball from each jar in sequence and called these "random primes."

He then showed by probability theory that there is an infinity of "twin random primes," n and $n + 1$ (unlike ordinary primes, consecutive numbers can turn out to be random primes), and he drew other conclusions that matched the actual or conjectured behavior of the ordinary primes.

David Hawkins has studied another kind of random prime. Start with the natural numbers,

2, 3, 4, 5, 6, 7, 8, 9, 10, 11, 12, 13, 14, 15,
16, 17, 18, 19, 20, 21, 22, 23, . . .

but instead of striking out every even number, strike out each number with probability one-half, for example, by tossing a coin when you reach the number and throwing it out if you toss heads. This will, on average, eliminate half the numbers. Suppose the numbers remaining are

3, 4, 5, 7, 10, 11, 13, 14, 17, 18, 19, 23, . . .

The first remaining number, 3, counts as the first random prime. Next, throw out every number with probability one-third. The remaining numbers could be

3, 5, 7, 10, 11, . . .

So 5 is the next "random" prime number. Now strike out each number greater than 5 with probability one-fifth, and so on. The result is a sequence of "random" primes.

What can be said about these "random" primes? Well, with probability 1, the **prime number theorem** is true: the number of random primes less than n tends to $n/\log n$ as n increases. There is also in each sequence of random primes, with probability 1, an infinity of "twin random primes," n and $n + 1$, and their number tends to $n/(\log n)^2$ as n tends to infinity. **Heuristic** arguments suggest the same formula for the actual twin primes. (Hawkins 1958)

magic squares

Rouse Ball in his *Mathematical Recreations and Essays* presented this magic square composed entirely of primes, created by the great English puzzlist Henry Ernest Dudeney.

67	1	43
13	37	61
31	73	7

(Dudeney 1917, problem 408) (Rouse Ball 1939, 211)

Martin Gardner once offered through his *Scientific American* column $100 for a 3-by-3 magic square all of whose entries are consecutive primes. Harry Nelson collected the prize by using the prime $p = 1480028171$ in the central cell, and around it the primes $p \pm 12$, ± 18, ± 30, and ± 42.

The magic sum in this magic square of sixteen consecutive primes is much smaller, 258:

37	53	89	79
83	61	67	47
97	71	59	31
41	73	43	101

(Giles Blanchette: Caldwell, *Prime Pages*)

Matijasevic and Hilbert's 10th problem

One of the most notorious of **Hilbert's 23 problems** was the 10th: does there exist a universal algorithm for solving Diophantine equations, that is, polynomial equations with integral coefficients and integral solutions?

It was solved over many years by Martin Davis, H. Putnam, and Julia Robinson, who took the first steps, and Yuri Matijasevic, who took the final step in 1970, using the divisibility properties of the **Fibonacci numbers**.

The solution was negative: no such algorithm exists, yet this negative conclusion had some very positive implications. Matijasevic's proof implies that there is a polynomial in twenty-three variables, whose positive values, only, for integer values of the variables, are the set of primes.

In 1976 J. P. Jones, D. Sato, H. Wada, and D. Wiens published their version, a polynomial in twenty-six variables which, very conveniently, can be written down using the twenty-six letters of the alphabet:

$$(k + 2)\{1 - [wz + h + j - q]^2 - [(gk + 2g + k + 1)(h + j) + h - z]^2 - [2n + p + q + z - e]^2 - [16(k + 1)^3(k + 2)(n + 1)^2 + 1 - f^2]^2 - [e^3(e + 2)(a + 1)^2 + 1 - o^2] - [(a^2 - 1)y^2 + 1 - x^2]^2 - [16r^2y^4(a^2 - 1) + 1 - u^2]^2 - [((a + u^2(u^2 - a))^2 - 1)(n + 4dy^2) + 1 - (x + cu)^2]^2 - [n + l + v - y]^2 - [(a^2 - 1)l^2 + 1 - m^2]^2 - [ai + k + 1 - l - i]^2 - [p + l(a - n - 1) + b(2an + 2a - n^2 - 2n - 2) - m]^2 - [q + y(a - p - 1) + s(2ap + 2a - p^2 - 2p - 2) - x]^2 - [z + pl(a - p) + t(2ap - p^2 - 1) - pm]^2\}$$

At first sight this is very complicated. At second sight it looks at worst like cheating, and at best like a trick, because it's of the form $(k + 2)(1 - A^2 - B^2 - C^2 \ldots)$ where A, B, and C and so on are themselves polynomials.

In other words, it will *only* have positive values when $A = B = C \ldots = 0$, when the values will be $k + 2$ (which will be prime), and we could just as well present it as this list of conditions:

$$0 = wz + h + j - q$$
$$0 = (gk + 2g + k + 1)(h + j) + h - z$$
$$0 = 2n + p + q + z - e$$
$$0 = (16k + 1)^3(k + 2)(n + 1)^2 + 1 - f^2$$
$$0 = e^3(e + 2)(a + 1)^2 + 1 - o^2$$
$$0 = (a^2 - 1)y^2 + 1 - x^2$$
$$0 = 16r^2y^4(a^2 - 1) + 1 - u^2$$
$$0 = ((a + u^2(u^2 - a))^2 - 1)(n + 4dy)^2 + 1 - (x + cu)^2$$
$$0 = n + l + v - y$$
$$0 = (a^2 - 1)l^2 + 1 - m^2$$
$$0 = ai + k + 1 - l - i$$
$$0 = p + l(a - n - 1) + b(2an + 2a - n^2 - 2n - 2) - m$$
$$0 = q + y(a - p - 1) + s(2ap + 2p - p^2 - 2p - 2) - x$$
$$0 = z + pl(a - p) + t(2ap - p^2 - 1) - pm$$

Matijasevic's construction can be applied to other sets of numbers, in fact to any set that is *recursively enumerable*, meaning roughly that it is generated by addition, multiplication, decisions based on in-equalities, and selecting an element from a list. For example, the set of **Mersenne primes** is the set of positive values of this polynomial, in thirteen variables, where the variables range over the non-negative integers.

Mersenne numbers and Mersenne primes

The friar Marin Mersenne (1588–1648), a member of the order of Minims, was a philosopher, mathematician, and scientist who wrote on music, mechanics, and optics, suggested to Christian Huygens that a pendulum might be used to measure time, and defended Christianity against Skepticism. He also occupied a unique role as a correspondent who linked together no fewer than seventy-eight philosophers and mathematicians across Europe, including Descartes, Fermat, the English philosopher Thomas Hobbes, Blaise Pascal and his father, Etienne, who met as Mersenne's guests, and others who corresponded through him—he was a kind of international post box—including Huygens, Torricelli, and Galileo.

Mersenne numbers

If $2^n - 1$ is prime, then n is prime, because if n had a prime factor p then $2^p - 1$ would divide $2^n - 1$. If $n = pq$, then

$$2^n - 1 = (2^p - 1)(2^{pq-p} + \ldots + 2^{3p} + 2^{2p} + 2^p + 1)$$

The converse, however, is not always true. If n is prime, $2^n - 1$ may be either prime or composite.

Mersenne famously speculated in his book *Cogitata Physico-Mathematica* (1644) that $2^n - 1$ is prime when $n = 2, 3, 5, 7, 13, 17, 19, 31, 67, 127,$ and 257 and no other n less than 257.

This is an extraordinary list, because numbers such as $2^{67} - 1$ are so large: how could Mersenne have possibly been confident that they were prime? No one in his day had any satisfactory method of deciding whether such large numbers were prime or composite.

Anyway, Mersenne's list contains several mistakes. The numbers M_{67} and M_{257} are composite, and the missing M_{61}, M_{89}, and M_{107} are prime—but these errors were not finally demonstrated until three hundred years later, by A. Ferrier in 1947.

hunting for Mersenne primes

- In 1536 Hudalricus Regius showed that $2^{11} - 1 = 2047 = 23 \cdot 89$ is not prime.
- In 1603 Pietro Cataldi correctly announced that M_{17} and M_{19} were prime, and falsely claimed for M_{23}, M_{29}, and M_{37}. Fermat corrected him on M_{23}, and M_{37} in 1640.
- In 1750 Euler, who had already corrected Cataldi on M_{29}, showed that M_{31} is prime. This was the largest known prime for over a hundred years, from 1772 to 1876.
- In 1876 Lucas concluded that

  $$M_{127} = 170141183460469231731687303715884105727$$

  is prime, though he never seemed entirely convinced by his own calculations. This was the record large prime until 1951, and the largest ever to be calculated by hand.
- In 1883 Pervouchine showed that M_{61} is prime, missing from Mersenne's list.
- Powers announced in 1911, in "The Tenth Perfect Number" (*American Mathematical Monthly* 18), that M_{89} is prime, also correcting Mersenne. He also showed that M_{107} is prime, in 1914.

- In 1903 F. N. Cole proved that M_{67}, on Mersenne's list, is composite.
- Mersenne's list was finally and completely checked only after World War II: Mersenne numbers are prime for $n = 2, 3, 5, 7, 13, 17, 19, 31, 61, 89, 107,$ and 127, up to the limit 257.

the coming of electronic computers

In 1952 Robinson proved that $M_{521}, M_{607}, M_{1279}, M_{2203},$ and M_{2281} are prime, the first successes using a modern computer. The advent of electronic computers has made the hunt for Mersenne primes more, not less, competitive. When Gillies of the University of Illinois found the 23rd Mersenne prime, $2^{11213} - 1$, the math department changed its postage meter to print "$2^{11213} - 1$ is prime" on every envelope.

On November 14, 1978, a press release announced, "Two 18-year-old youths, Laura Nickel and Curt Noll, have calculated the largest known prime number, 2 to the 21,701st power minus one, using a terminal at California State University, Hayward and hooking into the CYBER 174 in the Los Angeles area. Totaling 6,533 digits, the number was proved to be prime last October 30 after three years of diligent study and work interrupted at times by official and personal problems." The news made the front page of the *New York Times*.

Nickel and Noll got hooked on computer calculation in high school, and wrote their program with information from the local Cal State University math department. They ran their program for 440 hours when the CYBER 174 was not otherwise in use.

This was the 25th Mersenne prime. Noll went on to discover the 26th, M_{23209}, which has 13,395 digits, and later worked for Silicon Graphics, where most of the recent largest prime number records have been set, by David Slowinski, Paul Gage, and others. Noll is also a member of the EFF Cooperative Computing Awards Team, and one of the Amdahl Six who discovered $235235 \cdot 2^{70000} - 1$, a number of 21,078 digits and briefly the largest non-Mersenne prime known. Silicon Graphics takes part in **GIMPS**, the Great Internet Mersenne Prime Search.

Currently we know that M_p is prime for $p = 2, 3, 5, 7, 13, 17, 19, 31, 61, 89, 107, 127, 521, 607, 1279, 2203, 2281, 3217, 4253, 4423, 9689, 9941, 11213, 19937, 21701, 23209, 44497, 86243, 110503, 132049, 216091, 756839, 859433, 1257787, 1398269, 2976221, 3021377, 6972593, 13466917, 20996011,$ and 24036583. For all other p less than 6977600, M_p is composite. If p is less than 10412700, it is probably composite.

Because the tests for Mersenne primes are relatively simple, the **record** largest known prime number has almost always been a Mersenne prime.

n	p	no. of digits	year	discoverer
13–14	521, 607	157, 183	1952	R. M. Robinson
15–17	1279, 2203, 2281	386, 664, 687	1952	R. M. Robinson
18	3217	969	1957	H. Riesel
19–20	4253, 4423	1,281, 1,332	1961	Hurwitz and Selfridge
21–23	9689, 9941, 11213	2,917, 2,993, 3,376	1963	Gillies
24	19937	6,002	1971	Tuckerman
25	21701	6,533	1978	Curt Noll and Laura Nickel
26	23209	6,987	1979	Curt Noll
27	44497	13,395	1979	Slowinski and Nelson
28	86243	25,962	1982	Slowinski
29	132049	33,265	1983	Slowinski
30	216091	39,751	1985	Slowinski
31	110503	65,050	1988	Colquitt and Welsh
32	756839	227,832	1992	Slowinski and Gage
33	859433	258,716	1994	Slowinski and Gage
34	1257787	378,632	1996	Slowinski and Gage
35	1398269	420,921	1996	Armengaud, Woltman et al. (GIMPS team)
36	2976221	895,832	1996–97	Spence, Woltman et al. (GIMPS)
37	3021377	909,526	1998	Clarkson, Woltman, Kurowski et al. (GIMPS and PrimeNet)
38	6972593	2,098,960	1999	Hajratwala, Woltman, Kurowski et al. (GIMPS and PrimeNet)
??	13466917	4,053,946	2001	Cameron, Woltman, Kurowski et al. (GIMPS and PrimeNet)

| ?? | 20996011 | 6,320,430 | 2003 | Shafer, Woltman, Kurowski et al. (GIMPS and PrimeNet) |
| ?? | 24036583 | 7,235,733 | 2004 | Findley, Woltman, Kurowski et al. (GIMPS and PrimeNet) |

The latest Mersenne primes may possibly not be the 39th, 40th, and 41st, because not all the smaller Mersenne numbers have yet been checked.

This is how *MathWorld* Headline News, at http://mathworld.wolfram .com/news, reported (June 1, 2004) the latest record: "Josh Findley, a participant in the Great Internet Mersenne Prime Search (GIMPS), identified the 41st known Mersenne Prime on the morning of May 15. The discovery was confirmed on May 29, making $2^{24036583} - 1$ the largest known Mersenne prime, as well as the largest prime number known."

Mersenne prime conjectures

All known Mersenne primes are squarefree, and it seems likely that they all are, because it is known that if the square of a prime, p^2, divides a Mersenne prime, then p is a **Wieferich prime**, and these are very rare. Just two Wieferich primes less than $4 \cdot 10^{12}$ are known, and neither of them divides a Mersenne prime.

It was a natural speculation that if M_p is prime then M_{M_p} is also prime. For example, $M_2 = 3$, and sure enough, $M_3 = 7$ is prime: $M_5 = 31$ and M_{31} is prime, as Euler proved in 1750. But the conjecture is nevertheless wrong: $M_{13} = 8191$ is prime, but M_{8191} turns out to be composite.

Catalan in 1876 made a more limited conjecture: the sequence of Mersenne numbers

$$2^2 - 1 = 3 \qquad 2^3 - 1 = 7 \qquad 2^7 - 1 = 127$$

and $2^{127} - 1$ is also prime. Unfortunately, the next term has more than 10^{38} digits and cannot be tested at present.

the New Mersenne conjecture

This is the New Mersenne conjecture, made by Paul Bateman, Paul Selfridge, and Stan Wagstaff. Let p be an odd prime: then the following two statements are equivalent:

1. M_p is prime.
2. These two statements are either both true or both false:
   a. $(2^p + 1)/3$ is prime.
   b. p is of the form $2^k \pm 1$ or $4^k \pm 3$.

how many Mersenne primes?

It is not known whether there are infinity of Mersenne primes, or indeed an infinity of Mersenne composites. In *Unsolved Problems in Number Theory*, Richard K. Guy says of Mersenne primes, "their number is undoubtedly infinite, but proof is hopelessly beyond reach." He then offers some suggestions for the size of $M(x)$, the number of primes $p \leq x$ for which $2^p - 1$ is prime.

Gillies suggested $M(x)$ is approximately $c \log x$. The constant c could be e^γ, where γ is **Euler's constant**. Pomerance suggested $M(x)$ is approximately $c (\log \log x)^2$.

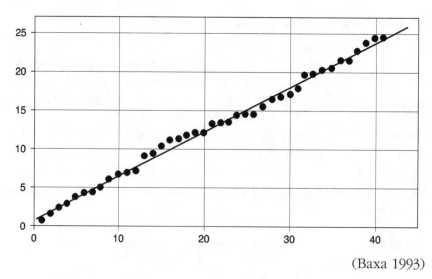

(Baxa 1993)

If the Mersenne primes, $2^p - 1$, are put on a graph by plotting $\log p$ against n, then you do get a remarkably "straight" line. On the basis of such data and an ingenious **heuristic** argument, Wagstaff conjectured in 1983 that:

1. M_x is about $(e^\gamma/\log 2)(\log \log x)$ or roughly $2.5695\ldots$ (log log x).

2. The number of Mersenne primes M_q with $x < q < 2x$ is about $e^\gamma = 1.7806\ldots$

3. The probability that M_q is prime is about $(e^\gamma \log aq)/(q \log 2)$ where $a = 2$ if q is of the form $4n + 3$ and $a = 6$ if q is of the form $4n + 1$.

Eberhart's conjecture

Eberhart conjectured that if q_n is the nth prime such that M_{q_n} is a Mersenne prime, then q_n is approximately $(3/2)^n$, if n is large enough.

factors of Mersenne Numbers

Proving that a Mersenne number is composite tells you nothing about its actual factors, and only some Mersenne composites have been factored. (Incidentally, Mersenne numbers written in base 2 consist of all 1s, and so they are binary **repunits** and there is a connection with the problem of factorizing the repunits in general.)

It helps to know some basic theorems about the factors of Mersenne primes. For example, if p is an odd prime, then any divisor of M_p is of the form $2kp + 1$.

Euler proved in 1750 that if p is a prime of the form $4n + 3$, then $2p + 1$ is a factor of M_p if and only if $2p + 1$ is also prime (so p and $2p + 1$ are a pair of **Sophie Germain primes**). In this case, M_p is composite, with the exception of the case $p = 3$, when $M_3 = 7$ and it is true that $2 \cdot 3 + 1 = 7$ divides M_3, but it is not composite.

For example, $M_{11} = 2047$, and 11 is prime and $11 = 4 \cdot 2 + 3$, and $2 \cdot 11 + 1 = 23$ is also prime, so $23 \mid 2047$, which is true.

Similarly, if n divides M_p, then n is of the form $8m \pm 1$. Using the same example, $M_{11} = 2047 = 23 \cdot 89$, and $89 = 8 \cdot 11 + 1$ and $23 = 8 \cdot 3 - 1$.

Here are just a handful of factorizations:

$$M_{11} = 2{,}047 = 23 \cdot 89$$
$$M_{23} = 8{,}388{,}607 = 47 \cdot 178{,}481$$

$$M_{29} = 233 \cdot 1,103 \cdot 2,089$$
$$M_{37} = 223 \cdot 616,318,177$$
$$M_{41} = 13,367 \cdot 164,511,353$$
$$M_{43} = 431 \cdot 9,719 \cdot 2,099,863$$
$$M_{47} = 2,351 \cdot 4,513 \cdot 13,264,529$$
$$M_{53} = 6,361 \cdot 69,431 \cdot 20,394,401$$
$$M_{59} = 179,951 \cdot 3,203,431,780,337$$
$$M_{67} = 193,707,721 \cdot 761,838,257,287$$
$$M_{71} = 228,479 \cdot 48,544,121 \cdot 212,885,833$$
$$M_{73} = 439 \cdot 2,298,041 \cdot 9,361,973,132,609$$
$$M_{79} = 2,687 \cdot 202,029,703 \cdot 1,113,491,139,767$$
$$M_{83} = 167 \cdot 57,912,614,113,275,649,087,721$$
$$M_{97} = 11,447 \cdot 13,842,607,235,828,485,645,766,393$$

As an example of what the latest methods can do, M_{619} was completely factored by the NFSNET team using the Special Number Field Sieve. Two factors were already known, 110183 and 710820995447. They have found these two remaining factors: 10937 8681671075 2971956924 8023421390 8123642560 1922510384 55204252439 and 253956 7680731642 1450129702 3118206917 3098610826 6999358245 0697816383 2424511153 6552907117 042045245 5686291833.

Lucas-Lehmer test for Mersenne primes

The Lucas-Lehmer test definitely determines whether a Mersenne number is prime or not.

The test was invented by Lucas and refined by Lehmer in about 1930. This is how it works. Define the sequence,

$$S(1) = 4, \ S(2) = 14, \ S(3) = 194, \ S(4) = 37634, \ldots$$

where $S(n + 1) = S(n)^2 - 2$. Then M_{2n+1} is prime if and only if M_{2n+1} divides $S(2n)$.

The values of $S(n)$ rapidly become very large indeed, but fortunately they do not need to be calculated directly to run the test. For example, $M_{13} = 8191$, so this will divide $S(12)$ if and only if 8191 is prime. Since we are only interested in divisibility, we can do the entire calculation modulo 8191, like this, using a hand calculator:

$S(4)$		$= 37634$		$\equiv 4870 \pmod{8191}$
$S(5)$	$\equiv 4870^2 - 2$	$= 23716898$		$\equiv 3953 \pmod{8191}$

$S(6)$	$\equiv$	$3953^2 - 2$	$= 15626207$	$\equiv 5970$	(mod 8191)
$S(7)$	$\equiv$	$5970^2 - 2$	$= 35640898$	$\equiv 1857$	(mod 8191)
$S(8)$	$\equiv$	$1857^2 - 2$	$= 3448447$	$\equiv 36$	(mod 8191)
$S(9)$	$\equiv$	$36^2 - 2$	$= 1294$	$\equiv 1294$	(mod 8191)
$S(10)$	$\equiv$	$1294^2 - 2$	$= 1674434$	$\equiv 3470$	(mod 8191)
$S(11)$	$\equiv$	$3470^2 - 2$	$= 12040898$	$\equiv 128$	(mod 8191)
$S(12)$	$\equiv$	$128^2 - 2$	$= 16382 = 8191 \cdot 2$	$\equiv 0$	(mod 8191)

And so 8191 is prime. Of course, this is not a brilliant way of testing a number the size of 8191 which we could also test by dividing by all the primes less than $\sqrt{8191} = 90.5\ldots$! But this method works efficiently for much larger numbers.

See GIMPS; primality testing

Mertens constant

The sum of the **reciprocals of the primes** diverges, but very slowly. Let

$$P(p) = \tfrac{1}{2} + \tfrac{1}{3} + \tfrac{1}{5} + \tfrac{1}{7} + \ldots + \tfrac{1}{p}$$

The sum from $\tfrac{1}{2}$ to $\tfrac{1}{p}$ is approximately log log p, and we have the beautiful result that $P(p) - \log \log p$ tends to the Mertens constant,

$$M = 0.261497212847642\ldots$$

as p tends to infinity. (Sloane A077761)

See Brun's constant

Mertens theorem

Approximately half of all numbers are divisible by 2, and so half are not. We have to say "approximately" because, for example, between 1 and 11 inclusive, only five out of eleven are even! (The argument only works perfectly for even numbers.)

Similarly, approximately one-half are not divisible by 3 and one-half multiplied by one-third are divisible by neither 2 nor 3. (This conclusion is also only exactly true for multiples of 2 and 3.) In general, approximately,

$$(1 - \tfrac{1}{2})(1 - \tfrac{1}{3})(1 - \tfrac{1}{5})(1 - \tfrac{1}{7}) \ldots (1 - \tfrac{1}{p_n})$$

are divisible by none of the primes from 2 to p_n. The table shows this product, approximately, for primes up to 1,000,000:

N	product of $(1 - \frac{1}{p_n})$ for primes $\leq N$
10	0.2286
10^2	0.1203
10^3	0.0810
10^4	0.0609
10^5	0.0488
10^6	0.0407

So about 77% of numbers are divisible by 2, 3, 5, or 7, while roughly 94% are divisible by a prime under 10,000. (Young 1998)

According to the **prime number theorem**, the number of primes $\leq n$ is $n/\log n$, and so the proportion of numbers $\leq n$ that are prime is $1/\log n$. Does this mean that the product

$$(1 - \tfrac{1}{2})(1 - \tfrac{1}{3})(1 - \tfrac{1}{5})(1 - \tfrac{1}{7}) \ldots (1 - \tfrac{1}{p_n})$$

tends to $1/\log n$ as n tends to infinity? No! The sources of error we have noted are cumulative, and the value of product tends to $e^{-\gamma}/\log n$, with an error factor roughly proportional to $\sqrt{n}$. This is Mertens theorem.

See heuristic reasoning

Mills' theorem

W. H. Mills proved in 1947 a theorem that is as amazing as it is useless. He proved that there is a number, A, such that A^{3^n} is prime for all integer values of n. Even more surprisingly, Mills did not give an actual value for A—his proof was nonconstructive.

This is indeed a **formula for prime numbers**, but it is useless in practice because you need to know A to a very large number of decimal places to get even a few primes out, and you need to know the primes generated before calculating A.

The smallest possible value of A, known as Mills' number, has been calculated as,

1.3063778838630806904686144926026057129167845851567 . . .

This leads to the sequence of primes starting, 2, 11, 1361, 2521008887, . . .

There is an infinity of other possible values of A.

Wright's theorem

Inspired by Mills, E. M. Wright proved that there is a number a, such that if $G_0 = a$ and $g_{n+1} = 2^{g_n}$ then $[g_n]$, the integral part of g_n, $n > 1$, is always a prime. He gave the example $a = 1.9287800\ldots$ when $[g_1] = 3$, $[g_2] = 13$, $[g_3] = 16381$, and $[g_4]$ has about 5,000 digits. (Wright 1951)

mixed bag

There is such a cornucopia of results about primes that they could not possibly fit into a much larger book than this, and nor can they all be easily and neatly classified, so here is a small bundle of miscellaneous results.

- Every sequence of seven consecutive numbers greater than 36 includes a multiple of a prime greater than 41. (Gupta: Caldwell, *Prime Pages*)
- Every number greater than 45 is the sum of distinct primes greater than 11. (Gupta: Caldwell, *Prime Pages*)
- Every number greater than 121 is the sum of distinct primes of the form $4n + 1$.
- The numbers 2, 5, 71, 369119, and 415074643 are the only known numbers that divide the sum of all the primes less than them.
- The number 1549 is the only odd number less than 10,000 that is not the sum of a prime and a power.
- The sequence of prime numbers embedded in the decimal expansion of pi (π) goes 3, 31, 314159 and then

    31,415,926,535,897,932,384,626,433,832,795,028,841

- The sequence of primes with consecutive digits starts: 23, 67, 89, 4567, 78901, 678901, 23456789, 45678901, . . . (*Journal of Recreational Mathematics* 5:254)
- The number 1683 is the only number N that can be expressed as a sum in exactly N ways using three distinct primes each time (the order of the primes in the sum is irrelevant). (Rivera: Caldwell, *Prime Pages*)
- There is a prime of the form $k \cdot 2^n + 1$ for each k less than 383. (Selfridge: Caldwell, *Prime Pages*)

- The number

$$82818079787776757473727171069686$$
$$76665646362616059585756555545352$$
$$51504948474645444342414039383 73$$
$$63534333231302928272625242322 21$$
$$2019181716151413121110987654321$$

is written by starting at 82 and working backwards to 1, and it is prime. (Nicol and Filaseta: Caldwell, *Prime Pages*)
- The largest prime below 1,000,000,000 is 999,999,937.
- The first triple of consecutive numbers each with three distinct prime factors is 1309-1310-1311. The next two such triples are 1885-1886-1887 and 2013-2014-2015.
- The first two pairs of consecutive numbers with four distinct prime factors each are 7314-7315, followed by 8294-8295 and 8645-8646.
- The set

$$5 \cdot 11 \cdot 13 - 2 \cdot 3 \cdot 7 \cdot 17 = 1$$

is the largest known set of consecutive primes that can be partitioned into two sets with this property. (Caldwell, *Prime Pages*)
- Every number greater than 55 is the sum of distinct primes of the form $4n + 3$.

multiplication, fast

Not only is it incomparably easier to multiply two large numbers together than it is to reverse the operation and factorize the product into the original numbers, but it is possible to multiply very large numbers almost as quickly as they can be added.

This seems counterintuitive. Any schoolchild will tell you that multiplication by hand is much more time-consuming than addition, and in general, following the usual algorithm for long multiplication, it would seem necessary to make about n^2 small calculations to multiply two n-digit numbers together (plus a much smaller number of additions).

Not so! Suppose that we have two numbers A and B represented in binary notation. If each has $2n$ digits, then we can split them into halves by writing them as

$$A = 2^n a + a' \quad B = 2^n b + b'$$
Then, $AB = (2^{2n} + 2^n)ab + 2^n(a - a')(b - b') + (2^n + 1)a'b'$

This has reduced the original multiplication—which we assumed would take $(2n)^2 = 4n^2$ multiplications plus a bit of addition—to three multiplications of about n^2 operations each.

This in itself is a saving of about 25%. However, when multiplying very large numbers, this process can be repeated. It turns out that the running time for large multiplications can be reduced from round about n^2 to $n^{1.585}$, a great saving.

Is this the limit? No! By even more cunning methods, extremely large numbers can be multiplied in a time that is nearly proportional to the length of the numbers. Even better, if you have a computer with an unlimited number of components all acting simultaneously, then it is possible to design a device such that, "If we wire together sufficiently many of these devices in a straight line, with each module communicating only with its left and right neighbours, the resulting circuitry will produce the $2n$-bit product of n-bit numbers in exactly $2n$ clock pulses." (Knuth 1981, 297) Now that is magic!

Niven numbers

Niven numbers are named after Ivan Niven, author of *An Introduction to the Theory of Numbers* (1960). (They were also labeled multidigital numbers, or Harshad numbers, by Kaprekar, *Harshad* meaning "great joy" in Sanskrit.) They are integers divisible by the sum of the digits. Their sequence in base 10 starts

1, 2, 3, 4, 5, 6, 7, 8, 9, 10, 12, 18, 20, . . .

(Sloane A005349)

In 1994, Grundman proved that at most twenty consecutive Niven numbers are possible and produced an infinite family of such sequences, the smallest one of 44,363,342,786 digits.

There is an infinity of numbers that are both Niven and **Smith**. (Weisstein, *MathWorld*)

odd numbers as $p + 2a^2$

The conjecture that every odd number can be written as $p + 2a^2$, where p is unity or a prime and a could be zero, is false. Only two counterexamples, however, are known, 5777 and 5993. (Ashbacher: Caldwell, *Prime Pages*)

Opperman's conjecture

In 1882 Opperman claimed that if $n > 1$ then $\pi(n^2 + n) > \pi(n^2) > \pi(n^2 - n)$. This has never been proved, though the evidence is compelling:

$n^2 - n$	primes in the gap	n^2	primes in the gap	$n^2 + n$
2	3	4	5	6
6	5	9	11	12
12	13	16	17, 19	20
20	23	25	29	30
30	31	36	37, 41	42
42	43, 47	49	53	56
56	59, 61	64	67, 71	72
72	73, 79	81	83, 89	90
90	93, 97	100	101, 103, 107, 109	110

. . .

See Brocard's conjecture

palindromic primes

There are fifteen palindromic primes consisting of three digits: 101, 131, 151, 181, 191, 313, 353, 373, 383, 727, 757, 787, 797, 919, and 929. The only palindromic prime with an even number of digits is 11.

The current record for the largest known palindromic prime is held by Harvey Dubner, who found on April 5, 2004, $10^{120016} + 1726271 \cdot 10^{60005} + 1$, which has 120,017 digits. His previous record was $(1989191989)_{1560}1$, containing 15,601 digits, announced in October 2001. The notation means that 1989191989 is repeated 1,560 times and then a 1 is put on the right-hand end.

Dubner also found, on May 4, 2004, this palindromic prime whose digit length, 98689, is also a palindrome:

$$10^{98689} - 429151924 \cdot 10^{49340} - 1$$

(www.worldofnumbers.com/palprim2)

pandigital primes

The digits 1 to 9 cannot be arranged to make a prime number, because their sum, 45, is divisible by 3, and so therefore is any arrangement of them. However, you can add an extra unit to make a prime. The first few of these are 10123457689, 10123465789, 10123465897, 10123485679, 10123485769, . . . (Sloane A050288)

If you exclude zero, the smallest pandigital primes are 1123465789, 1123465879, 1123468597, 1123469587, 1123478659, . . . (Sloane A050290)

Dubner and Ondrejka discovered the smallest pandigital prime that is also palindromic: 1023456987896543201 (Caldwell, *Prime Pages*)

Pascal's triangle and the binomial coefficients

```
                            1
                        1       1
                    1       2       1
                1       3       3       1
            1       4       6       4       1
        1       5      10      10       5       1
      1     6     15     20     15      6      1
    1     7     21     35     35     21      7      1
  1     8    28    56    70    56    28     8     1
 1    9    36    84   126   126    84    36    9    1
1   10   45   120  210  252  210  120   45   10   1
1  11  55  165  330  462  462  330  165  55  11  1
```

Pascal's triangle is formed by starting and ending every row with 1, and using the rule that each entry is the sum of the two entries above it in the previous row.

The entries are also the *binomial coefficients* that appear when calculating algebraic expressions such as

$$(1 + n)^8 = 1 + 8n + 28n^2 + 56n^3 + 70n^4 + 56n^5 + 28n^6 + 8n^7 + n^8$$

Binomial coefficients can also be calculated directly. Counting the unit at the top of the triangle as row zero, the 4th entry in the 8th row is 56, and

$$\frac{8 \cdot 7 \cdot 6}{1 \cdot 2 \cdot 3} = 56$$

Ignoring the end units, which entries are divisible by their row number if we count the top unit as row zero? A quick check shows that if p is prime, then all of row p (the units excluded) are divisible by p.

This is the pattern created. Zero (0) indicates an entry that is divisible by the row number without remainder:

```
                              -
                      -   0   -
                  -   0   0   -
              -   0   -   0   -
          -   0   0   0   0   -
      -   0   -   -   -   0   -
  -   0   0   0   0   0   0   -
-   0   -   0   -   0   -   0   -
- 0   0   -   0   0   -   0   0   -
- 0   -   0   0   -   0   0   -   0   -
- 0   0   0   0   0   0   0   0   0   0   -
```

It is less obvious, but true, that the entries in row n, excluding the ends, are divisible by prime p if and only if n is a power of p.

The middle term in every other row is $\binom{2n}{n}$, which is always divisible exactly by $n + 1$.

How many coefficients are not divisible by the row number? Surprisingly, the number as a proportion of all the entries drops to zero as the number of rows tends to infinity. *Almost all* entries are divisible by their row number. (Harborth 1977)

Pascal's triangle and Sierpinski's gasket

There is an obvious similarity between these figures and Sierpinski's gasket, which is constructed by deleting the central quarter of the triangle, then the central quarters of each of the three remaining triangles, and so on:

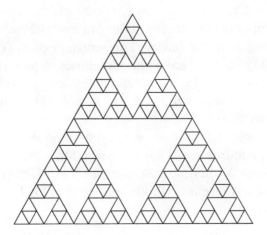

The Fibonacci sequence also appears in Pascal's triangle, along these diagonals:

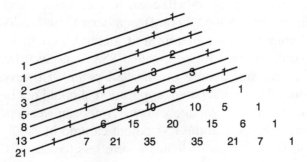

Pascal triangle curiosities

- How many times does a particular number occur in Pascal's triangle? David Singmaster pointed out that 120, 210, 1540, 7140, 11628, and 24310 each occur six times, among numbers less than 2^{48}. He has also proved that an infinity of numbers occur at least six times each.

- Erdös conjectured in the 1930s that if $n > 4$, then $\binom{2n}{n}$ is never **squarefree**, and Sarkösy proved it in 1985 for all sufficiently large n. Goetgheluck has confirmed that it is true if n is not a power of 2 and if $n < 2^{42205184}$.

- Erdös, Lacampagne, and Selfridge have conjectured that if p is the smallest prime factor of the binomial coefficient $\binom{N}{k}$, and k is at least 2, then p is at most $\binom{N}{k}$ or 29 whichever is the larger.

- The number of odd entries in the nth row is 2^k, where k is the number of 1s in the binary representation of n. For example, $11 = 1011_2$, so there are eight odd entries in row 11. In particular, in rows 3, 7, 15, 31, . . . all the entries are odd.

- For positive integers n and k and $n \geq 2k$, $\binom{n}{k}$ has a prime factor greater than k.

- Erdös proved that at least one of the binomial coefficients $\binom{n}{r}$ is not square-free if $n > 23$. Moreover, for some N_r, depending on r, if $n > N_r$ then at least one of $\binom{n}{r}$ is divisible by an rth power of some prime. (Erdös and Eynden 1992)

patents on prime numbers

The best-known use of number theory in the real world is the use of primes in **public key encryption**. This makes certain large primes and algorithms commercially valuable. So Rivest, Shamir, and Adleman filed patent number 4,405,829 on December 14, 1977, titled "Cryptographic Communications System and Method," on their **RSA algorithm**. A company was formed, called RSA Data Security, and was granted an exclusive license on the RSA patent. However, since it was published before the patent application, it could not be patented under European and Japanese law. American readers may be encouraged to know that the RSA patent expired in 2000!

Ralph Merkle and Martin Hellman were less fortunate. On August 19, 1980, they filed a patent, No. 4,218,582, for a "Public-Key Cryptographic Apparatus and Method," based on the knapsack problem, which requires items of different size to be fitted into a knapsack of given volume. Two years later Leonard Adleman solved the simplest version of Merkle's problem. Merkle promptly advertised in the pages of *Time* magazine a reward of $1,000 for anyone who could break a repeated iteration knapsack problem. Again he failed: Ernie Brickell cracked an iteration problem with forty iterations in just one hour on a Cray-1 supercomputer. That was the end of Merkle's knapsack method for all practical purposes.

More recently, Roger Schlafly obtained U.S. Patent 5,373,560 in 1994 for two primes, of which this is the smaller, represented in hexadecimal notation:

> 98A3DF52AEAE9799325CB258D767EBD1F4630E9B9E217
> 32A4AFB1624BA6DF911466AD8DA960586F4A0D5E3C36
> AF099660BDDC1577E54A9F402334433ACB14BCB

Since Schlafly is a member of the League for Programming Freedom, which is opposed to patenting computer software, he presumably had his tongue in his cheek.

See RSA algorithm

Pépin's test for Fermat numbers

Father Jean Pépin (1826–1904) was a Jesuit priest who published several papers on methods of factorization. In 1877 he published his test for the primality of **Fermat numbers**.

$F(k)$ is prime if and only if $k > 1$ and $5^{(F(k)-1)/2} + 1$ is divisible by $F(k)$
For example, $F(2) = 17$ and $5^{(17-1)/2} + 1 = 5^8 + 1 = 390626 = 17 \cdot 22978$

The test is usually presented today with 3 in place of 5.

Ironically, however, because so few Fermat primes are known, Pépin's test has never been used successfully to show that a Fermat number is prime!

In 1905 J. C. Moorhead and A. E. Western both used Pépin's test to show that F_7 is composite without finding the factors, which were calculated in 1970 by John Brillhart and M. Morrison:

$$F_7 = (2^9 \cdot 116{,}503{,}103{,}764{,}643 + 1)$$
$$\cdot (2^9 \cdot 11{,}141{,}971{,}095{,}088{,}142{,}685 + 1)$$

In 1909 Moorhead and Western collaborated to prove that F_8 is composite, using Pépin's test. (Koshy 2002)

The largest Fermat number that has been tested using Pépin's theorem is F_{24} in 1999 by Mayer, Papadopoulus, and Crandall. (Rivera, Conjecture 4)

perfect numbers

> It is recorded that all God's works were completed in six days, because six is a perfect number.... For this is the first number made of divisors, a sixth, a third, and a half, respectively, one, two, and three, totaling six.
>
> —*St. Augustine of Hippo,* The City of God

A perfect number is the sum of all its proper divisors (so it is neither **abundant** nor **deficient**). The second perfect number is 28, which Plutarch noted was the number of members in the Senate at Sparta.

On the other hand, Alcuin (735–804), who was Charlemagne's teacher, noted that since the human race was descended from eight souls in Noah's Ark, and 8 is deficient, therefore this second creation was inferior to the first. (Ore 1948/1988, 94)

Euclid presented perfect numbers in his *Elements*. He defined them in VII, Definition 22, and then in Book IX proposition 36, claimed that:

> If as many numbers as we please beginning from a unit be set out continuously in double proportion, until the sum of all becomes prime, and if the sum multiplied into the last make some number, the product will be perfect.

In other words, you add up the sequence 1, 2, 4, 8, 16, . . . until you get a prime and then multiply the sum by the last term. So the first four perfect numbers are:

$$6 = (1 + 2) \cdot 2$$
$$28 = (1 + 2 + 4) \cdot 4$$
$$496 = (1 + 2 + 4 + 8 + 16) \cdot 16$$
$$8128 = (1 + 2 + 4 + 8 + 16 + 32 + 64) \cdot 64$$

and sure enough, each number is the sum of its proper factors:

$$6 = 1 + 2 + 3$$
$$28 = 1 + 2 + 4 + 7 + 14$$
$$496 = 1 + 2 + 4 + 8 + 16 + 31 + 62 + 124 + 248$$
$$8128 = 1 + 2 + 4 + 8 + 16 + 32 + 64 + 127 + 254$$
$$+ 508 + 1016 + 2032 + 4064$$

Euclid claimed that every number of this form is perfect, but not that every (even) perfect number is of this form: that was proved by Euler in 1747. It follows that finding perfect numbers amounts to finding

Mersenne primes and the largest known perfect number will be derived directly from the record Mersenne prime.

The first four perfect numbers were known to Nicomachus of Gerasa (c. AD 100), and to Iamblichus, who wrote a commentary on Nicomachus's *Arithmetic*, and suggested that there might be one perfect number for each number of digits, suggesting that there is something special about base 10. This is false.

It was also a natural assumption that since 2, 3, 5, and 7 produce perfect numbers, the next would be for 11, which is also false: $2^{11} - 1 = 2047 = 23 \cdot 89$.

The perfect numbers also seem to end in either 6 or 8, which is true, and to do so alternately, which is again false. The final digits of the known perfect numbers go, 6, 8, 6, 8, 6, 6, 8, 8, 6, 6, 8, 8, 6, 8, 8, 8, 6, 6, 6, 8, 6, 6, 6, 6, 6, 6, 6, 8, 8, 6, 8, 8, 6, 8, 6, 6, . . . (Sloane A094540)

The fifth perfect number was found in 1536 by Hudalricus Regius, who showed that $2^{13} - 1 = 8191$ is prime, so $33550336 = 2^{12}(2^{13} - 1)$ is perfect. Then J. Schleybl in 1555 found the sixth: $8589869056 = 2^{16}(2^{17} - 1)$. Pietro Antonio Cataldi (1548–1626) rediscovered the fifth and sixth and added the seventh: $137438691328 = 2^{18}(2^{19} - 1)$. Cataldi, like Mersenne later, then made an error. He correctly claimed that $2^p - 1$ is prime for $p = 2, 3, 5, 7, 13, 17,$ and 19 and then added 23, 29, 31, and 37, of which only 31 is correct.

Peter Barlow was a very reliable calculator whose *Tables* (1814) of factors, squares, cubes, square roots, reciprocals, and natural logarithms up to 10,000 is still in print. He made a famous remark in his book *Elementary Investigation of the Theory of Numbers* (1811) that $2^{30}(2^{31} - 1) = 2,147,483,647$ "is the greatest [perfect number] that will be discovered; for as they are merely curious, without being useful it is not likely that any person will attempt to find one beyond it."

How close can a non-perfect number be to perfection? Powers of 2 are only just **deficient**: so if $n = 2^k$, then $\sigma(n) = 2n - 1$. Such numbers are called *almost perfect*. It isn't known if there are any others.

What about $\sigma(n) = 2n + 1$? These have been called quasi-perfect, and they must be odd squares, but no one knows if any exist. (Guy)

Every multiple of a perfect number is **abundant**, and every factor is **deficient**.

odd perfect numbers

Descartes, one of Mersenne's many correspondents, wrote to him in 1638 that:

> I think I am able to prove that there are no even numbers which are perfect apart from those of Euclid; and that there are no odd perfect numbers, unless they are composed of a single prime number, multiplied by a square whose root is composed of several other prime numbers. But I can see nothing which would prevent one from finding numbers of this sort. For example, if 22021 were prime, in multiplying it by 9018009 which is a square whose root is composed of the prime numbers 3, 7, 11, 13, one would have 198585576189, which would be a perfect number. But, whatever method one might use, it would require a great deal of time to look for these numbers.

Descartes points out that if $p = 22021$ were prime and if $s = 3 \cdot 7 \cdot 11 \cdot 13$, then ps^2 would be an *odd perfect number*—but $22021 = 61 \cdot 19^2$. (Crubellier and Sip 1997, 389–410)

All even perfect numbers are of Euclid's form, but could there be an odd perfect number? No one knows, but it is a typical curiosity of mathematics that you can say a great deal about an object that may not exist at all! So mathematicians have proved that *if* an odd perfect exists, it has at least eleven prime factors, must be greater than 10^{300}, the largest prime factor must be greater than 500,000, the second largest must be greater than 1000, and its prime factorization must consist of even powers of all its prime factors but one, which appears as an odd power.

On the other hand, if it has k distinct prime factors, then it is less than 4^{4^k}. (Guy 1994)

The only even perfect number of the form $a^n + b^n$ with $n > 1$, and a and b coprime, is 28.

See perfect, multiply; Sierpinski numbers; unitary perfect

perfect, multiply

A number is multiply perfect if $\sigma(n) = kn$, with $k \geq 2$, so ordinary perfect numbers are doubly perfect. The first triply perfect number (P_3) for which $\sigma(n) = 3n$ is $120 = 2^3 3 \cdot 5$: $\sigma(120) = 360$.

Fermat found the second triply perfect number, $672 = 2^5 3 \cdot 7$ for which $\sigma(672) = 2016 = 3 \cdot 672$. The third P_3, $523776 = 2^9 \cdot 11 \cdot 31$, was discovered by Father Jumeau in 1638.

Descartes announced this P_3: $1476304896 = 2^{13}3 \cdot 11 \cdot 31$

and these six P_4s:

30240	$= 2^5 3^3 5 \cdot 7$
32760	$= 2^3 3^2 5 \cdot 7 \cdot 13$
23569920	$= 2^9 3^3 5 \cdot 11 \cdot 31$
142990848	$= 2^9 3^2 7 \cdot 11 \cdot 13 \cdot 31$
66433720320	$= 2^{13} 3^3 5 \cdot 11 \cdot 43 \cdot 127$
403031236608	$= 2^{13} 3^2 7 \cdot 11 \cdot 13 \cdot 43 \cdot 127$

and also this P_5: $14182439040 = 2^7 3^4 \cdot 5 \cdot 7 \cdot 11^2 \cdot 17 \cdot 19.$

The common prime factors in these examples are due to Descartes's method for creating new multiply perfect numbers from old, which included these rules:

1. If n is a P_3 not divisible by 3, then $3n$ is a P_4.
2. If a P_3 is divisible by 3, but by neither 5 nor 9, then $4 \cdot 5 \, P_3$ is a P_4.
3. If a P_3 is divisible by 3, but not by 7, 9, or 13, then $3 \cdot 7 \cdot 13 \, P_3$ is a P_4.

(Dickson 1952, vol. 1, 33–35)

D. N. Lehmer proved that a P_3 must have at least three distinct prime factors, a P_4 at least four, a P_5 at least six, a P_6 at least nine, and a P_7 at least fourteen.

It is possible to create higher-order perfect numbers from lower-order, based on this theorem: if p is prime, n is p-perfect and $p \nmid n$, then pn is $(p + 1)$-perfect. For example, if n is 3-perfect and $3 \nmid n$, then $3n$ is 4-perfect, and if n is 5-perfect and $5 \nmid n$, then $5n$ is 6-perfect.

The number of known multiperfect numbers has risen steadily:

year	1911	1929	1954	1992	2004
no.	251	334	539	c. 700	5,040

The numbers of known n-multiperfect numbers are 1, 37, 6, 36, 65, 245, 516, 1134, 2036, 644, 1, 0, . . . including, it is believed, all those with index 3 to 7. (Weisstein, *MathWorld*) The current records are:

9-perfect	7.9842491755534198 . . . $\cdot\ 10^{465}$	1992	F. W. Helenius
10-perfect	2.86879876441793479 . . . $\cdot\ 10^{923}$	1997	Ron M. Sorli
11-perfect	2.51850413483992918 . . . $\cdot\ 10^{1906}$	2001	G. F. Woltman

None of these are claimed to be the smallest possible for their index. (Achim Flammenkamp, www.uni-bielefeld.de)

The Sloane sequences for the multiperfect numbers start:

- 3-perfect (Sloane A005820): 120, 672, 523776, 459818240, 1476304896, 51001180160, . . .
- 4-perfect (Sloane A027687): 30240, 32760, 2178540, 23569920, 45532800, 142990848, 1379454720, 43861478400, . . .
- 5-perfect (Sloane A046060): 14182439040, 31998395520, 518666803200, 13661860101120, 30823866178560, 740344994887680, 796928461056000, . . .
- 6-multiperfect (Sloane A046061): 154345556085770649600, 9186050031556349952000, 680489641226538823680000, 6205958672455589512937472000, . . .

See pseudoperfect numbers; weird numbers

permutable primes

Permutable primes remain prime when their digits are jumbled. Permutable primes are also **circular primes**, and like circular primes they are likely to be only finite in number.

Those with not more than 466 digits are (giving the smallest permutation only):

$$13, 17, 37, 79, 113, 199, 337$$

π, primes in the decimal expansion of

If p_k denotes the integer given by the first k decimal digits of π, then p_k is prime for $k = 1, 2, 6,$ and 38. In other words, these numbers are prime:

$$p_1 = 3$$
$$p_2 = 31$$
$$p_6 = 314159$$
$$p_{38} = 31415926535897932384626433832795028841$$

(Sloane A005042)

Edward Prothro has calculated (December 2001) that p_{500} to p_{16207} are all composite, and the next *probable* prime is p_{16208}, which passes the strong pseudoprime test, so the chance that it is composite is extremely small. The next prime p_k can be expected round about $k = 561460$.

See pseudoprimes, strong

Pocklington's theorem

Henry Cabourn Pocklington (1870–1952) was a physicist and a Fellow of the Royal Society. His contribution to mathematics was a single paper written during the First World War. D. H. Lehmer in 1927 used this paper to produce Pocklington's test for primality. Whereas **Pépin's test** for the primality of n depends on the complete factorization of $n - 1$, Pocklington's needs only a partial factorization.

Suppose that we can factorize $n - 1$ into two factors, $n - 1 = FR$ where F and R are coprime and $F > R$. Then if for every prime factor q of F there is an integer $a > 1$ such that,

$$a^{n-1} \equiv 1 \pmod{n},$$

and also the greatest common divisor of $a^{(n-1)/q} - 1$ and n is equal to 1, then n is prime. (There is no need to use the same value of a for every factor q.)

Polignac's conjectures

In 1849, Alphonse de Polignac (1817–1890) conjectured that for every even number $2k$ there is an infinity of primes that differ by $2k$. If $k = 1$ this reduces to the **twin primes** conjecture. (Dickson 1952, 424) (Polignac 1849)

See Dickson's conjecture

Polignac or obstinate numbers

Polignac also conjectured that every odd number is the sum of a prime and power of 2. He claimed to have verified this up to 3,000,000, and if he did indeed go to these lengths, it was a monumental waste of effort, for Euler had already noted that 127 and 959 cannot be so. In fact, Erdös proved that there is an infinite arithmetic progression of odd integers that cannot be represented as sum of a prime and 2^n.

The numbers that refute Polignac's conjecture have been called Polignac numbers or *obstinate* numbers, of which there are seventeen less than 1000, only one of which is composite, 905. (Ajax: Caldwell, *Prime Pages*) The sequence of failures starts: 127, 149, 251, 331, 337, . . .

Curiously, as Clifford Pickover has noted, there are many pairs of obstinate numbers differing by 2, such as 905 and 907; 3341 and 3343; 3431 and 3433. Pickover has also calculated the smallest titanic obstinate number, $10^{999} + 18919$.

If you want to join Pickover in finding obstinate numbers, you can sign up for the Grand Internet Obstinate Number Search, another example of **distributed computing**.

Most Polignac numbers are themselves prime. Cohen and Selfridge showed that the twenty-six-digit number,

$$47,867,742,232,066,880,047,611,079$$

is prime and neither the sum nor the difference of a power of 2 and a prime.

powerful numbers

If when a positive integer is factorized, all its prime factors appear at least squared, then it is *powerful*. The sequence of powerful numbers starts

$$4, 8, 9, 16, 25, 27, 32, 36, 49, 64, 72, 81, 100, . . .$$

Solomon Golomb showed how to find an infinite number of consecutive pairs of powerful numbers, such as 8 and 9, 288 and 289, and proved that the number of powerful numbers $\leq x$ is approximately $c\sqrt{x}$ where $c = 2.173$ . . . He also conjectured that 6 is not the difference between two powerful numbers, but this is false:

$$6 = 5^4 7^3 - 463^2$$

(Narkiewicz: Guy 1994)

It turns out that every integer is the difference between two powerful numbers.

The pairs of consecutive powerful numbers already mentioned are related to solutions to the Pell equation, $x^2 - 2y^2 = \pm 1$. If x and y satisfy this equation, then $8x^2y^2$ and $(x^2 + 2y^2)^2$ are consecutive powerful numbers.

x	y	$8x^2y^2$	$(x^2 + 2y^2)^2$
1	1	$8 = 2^3$	$9 = 3^2$
3	2	$288 = 2^5 \cdot 3^2$	$289 = 17^2$
7	5	$9800 = 2^3 \cdot 5^2 \cdot 7^2$	$9801 = 3^4 \cdot 11^2$
17	12	$332928 = 2^5 \cdot 3^2 \cdot 17^2$	$332929 = 577^2$

The rule for forming each new solution is that $x_{n+1} = x_n + 2y_n$ and $y_{n+1} = x_n + y_n$.

As a more general rule, if A and B are consecutive powerful numbers, then so are $4AB$ and $4AB + 1$. Also, if u and $u + 4$ are both powerful, then so are $u(u + 4)$ and $u(u + 4) + 4 = (u + 2)^2$.

Almost all known pairs of consecutive powerful numbers have one of them a perfect square. Golomb gives this exception: $23^3 = 12167$ and $2^3 \cdot 3^2 \cdot 13^2 = 12168$. (Golomb 1970)

Also, every sufficiently large number is the sum of three powerful numbers. (Heath-Brown: Guy 1994)

It has been conjectured that there cannot be three consecutive powerful numbers. (Golomb 1970)

The smaller number of each pair of consecutive powerful numbers is 8, 288, 675, 9800, 12167, 235224, . . . (Sloane A060355)
    Mollin and Walsh actually proved a much stronger result: for every integer n, there is an infinite number of pairs of powerful numbers whose difference is n. Also, every even integer is the difference in an infinity of ways of two powerful numbers neither of which is a square. (Mollin and Walsh 1986)
    See abc conjecture

primality testing

The problem of distinguishing prime numbers from composite numbers and of resolving the latter into their prime factors is

> known to be one of the most important and useful in arithmetic.
> It has engaged the industry and wisdom of ancient and modern
> geometers to such an extent that it would be superfluous to dis-
> cuss the problem at length. . . . Further, the dignity of the science
> itself seems to require that every possible means be explored for
> the solution of a problem so elegant and so celebrated.
>
> —*Gauss*, Disquisitiones Arithmeticae *(1801), #329*

Gauss went on to claim that with his methods "numbers with seven,
eight, or even more digits have been handled with success and
speed beyond expectation," though he admitted that "all methods
that have been proposed thus far are either restricted to very special
cases or are so laborious and prolix that . . . these methods do not
apply at all to larger numbers."

The problems of primality testing and factorization are indeed extraordinarily tricky and subtle. Fortunately, testing for primality is the
easier of the two.

Using powerful electronic computers, the sieve of Erastosthenes is
efficient for numbers up to about 10^{10}, while trial division, dividing a
number n by the primes, starting with 2, 3, 5, 7, and so on up to the
limit of $\sqrt{n}$, works for numbers with up to about twenty-five digits,
and both these methods will produce the factors, if there are any: primality simply means no proper factors.

To test for primality alone, however, there are more powerful methods. What is needed is a theorem about prime numbers that can be
turned around to become a test. One candidate is **Wilson's theorem**
that if and only if p is prime,

$$p \mid (p - 1)! + 1$$

However, calculating $(p - 1)!$ is extremely expensive even on large
modern computers, so this theorem is more or less useless. A more
promising candidate is **Fermat's Little Theorem**, which says if p is
prime and a and p are coprime then,

$$a^{p-1} \equiv 1 \pmod{p}$$

Unfortunately, there is a fly in the ointment!—the converse of Fermat's
Little Theorem is not true. There are **pseudoprimes**, composite numbers, q, for which the theorem is true for certain values of a (and there

are even absolute pseudoprimes, or **Carmichael numbers**, for which it is true for *any* value of *a*). These pseudoprimes are annoying, although they are relatively rare. For example, there are more than 10^9 primes less than $25 \cdot 10^9$ but only 21,853 base 2 pseudoprimes.

So Fermat's Little Theorem can be used as a test for primality, but not as a proof. If a number fails Fermat's Little Theorem, then it must be composite. But if it passes the test, there remains a (small) probability that it is not a real prime but only a pseudoprime. Hence a number that has passed the Fermat's Little Theorem test, maybe for several values of *a*, is called a *probable prime*, and the numbers *a* are its witnesses.

Suppose that you choose an odd number, *n*, at random so that $1 < n \leq x$, and a base *a* at random, $1 < a < n - 1$, and suppose that $a^{n-1} \equiv 1 \pmod{n}$ so that *n* is a pseudoprime to base *a*. What is the chance that *n* is actually composite, as a function of *x*?

Kim and Pomerance (1989) showed that if *x* has 60 digits, the chance is less than 0.0716, small but significant. However, if *x* has 100 digits, it is less than 0.0000000277, and if *x* has 1,000 digits, it is $1.6 \cdot 10^{-1331}$. Pseudoprimes are a nuisance, but as the range of *x* increases they become less and less significant.

There are several ways to get around these difficulties, leading to the **Lucas-Lehmer test**, the special Lucas-Lehmer test for **Mersenne numbers**, **Pépin's test for Fermat numbers**, **Pocklington's theorem**, **Proth's theorem**, methods based on **elliptic curves**, not forgetting the **AKS algorithm**, and newer and far more powerful algorithms, many of them *probabilistic*.

probabilistic methods

Probabilistic primality tests are similar to Fermat's Little Theorem, in that they search for *witnesses* (corresponding to the base in Fermat's

---

Titanic Primes

Samuel Yates defined a titanic prime as a prime with at least a thousand digits in 1984 when only 110 such primes had been identified. A titan, naturally, was anyone who had found a titanic prime. The name has now been made absurd by advances in computer hardware and programming methods!

theorem) to the compositeness of a number, N. If a witness is found, the algorithm will stop, and the conclusion is certain that the number is composite.

So the strategy is to choose some numbers, say, 200 of them, at random between 1 and $N - 1$ and test if they are witnesses. If they all are, and the chance of the number not being composite is one-half for each witness, then you can be confident that N is *not* prime, apart from the miniscule probability of 1 in 2^{200} that you are wrong. This is certainly close enough to certainty for so-called industrial grade primes for use in RSA encryption, for example.

For an introduction to more detailed and advanced arguments, an excellent source is Hugh Williams, *Édouard Lucas and Primality Testing* (1998).

prime number graph

The prime numbers can be illustrated by marking the points (n, p_n) on a graph:

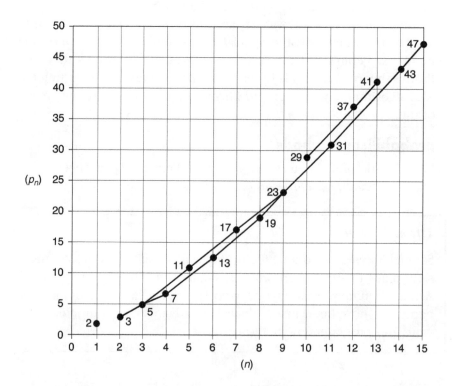

Carl Pomerance has proved that for every number n, there are n points on the prime number graph that are collinear. The illustration shows the first cases for $n = 3$, 4, and 5. (Pomerance 1979)

prime number theorem and the prime counting function

The questions, "How many primes are there?" and "How big is the nth prime?" are closely related. If we could answer either of them exactly with a simple formula, then we could also answer the other—but we can't. The best mathematicians have managed is to get more or less close to the exact answers.

The number of primes less than or equal to n is denoted by $\pi(n)$. This shouldn't be misleading—there is no connection between the π in $\pi(n)$ and the constant π that is related to the circle.

Since all primes except 2 and 3 are of the form $6n \pm 1$, we can say that at most one-third of numbers are prime, but this is a gross over-estimate. Noticing that they are also of the form $30n \pm 1$, 7, 11, or 13 means that at most eight out of thirty, or 26.66%, can be prime, but this is also a feeble figure, not least because the number of primes decreases the further we go but this proportion stays the same.

The prime number theorem asserts that as n increases, $\pi(n)$ *asymptotically approaches* $n/\log n$, meaning that $\pi(n) \, / \, n/\log n$ tends to 1 as n tends to infinity. In notation: $\pi(n) \sim n/\log n$. This is a brilliantly simple estimate, and a pretty good one, though not the best: it also turned out to be very difficult to prove.

The approximation $n/\log n$ to $\pi(n)$ also means that of the numbers 1 to n, roughly $1/\log n$ are prime. We could also say that the probability of a random number between 1 and n being prime is approximately $1/\log n$.

It also means that the average gap between two consecutive primes near the number x is close to $\log x$. Thus, when x is round about 100, $\log x$ is approximately 4.6, so roughly every fifth number should be prime.

history

Legendre was the first to put a version of the prime number theorem into print. In 1798 he claimed in his book *Essai sur la Théorie des*

Nombres that $\pi(x)$ is approximately $x/(\log x - 1.08366)$. He must have been tempted to conclude that the last figure should be 1 (the best value), but he stuck with his data, which only went up to 400,000.

The youthful Gauss had also been studying prime numbers, and in 1792 at the age of fifteen he proposed the estimate,

$$\pi(x) \text{ is approximately } \text{Li}(x) = \int_2^x \frac{1}{\log t}\, dt$$

Gauss's conjecture is equivalent to the prime number theorem.

There was then a long pause until Tchebycheff proved in 1851 that if $\pi(x) / x/\log x$ does have a limit, then the limit must be 1, though he was unable to take the final step and show that the limit existed. A year later he proved that if n is large enough, then

$$\frac{(0.92\ldots)x}{\log x} < \pi(x) < \frac{(1.105\ldots)x}{\log x}$$

The final step was taken entirely independently by de la Vallée Poussin (1866–1962) and Jacques Hadamard (1865–1963), coincidentally two of the longest lived mathematicians ever. They proved that, indeed,

$$\pi(x) \sim \frac{x}{\log x}$$

The error in this approximation depends on the zero-free region of the Riemann zeta function inside the critical strip within which the real part of x is between 0 and 1. The more we know about this region, the smaller we can make the error term. Koch showed in 1901 that if the **Riemann hypothesis** is true, then

$$\pi(x) = \text{Li}(x) + \text{ an error term of the order of } x^{1/2\,\log x}$$

De la Vallée Poussin also showed that Gauss's estimate $\text{Li}(x)$ is a better approximation to $\pi(x)$ than $x/(\log x - a)$ no matter what value is assigned to the constant a (and also that the best value for a is 1).

elementary proof

It was long believed that the prime number theorem could only be proved by analytic methods, using calculus. However, in 1949 Paul

Erdös and Atle Selberg found a method of proof that did not use the Riemann zeta function or the theory of complex numbers, and so is described as "elementary," though it is more complicated and harder to understand than the analytic proofs. Donald Benson recalls that in 1950 he attended a course of thirty lectures devoted entirely to explaining the new "elementary" proof! (Benson 1999, 231)

record calculations

The record value for $\pi(x)$ is $\pi(10^{21}) = 21{,}127{,}269{,}486{,}018{,}731{,}928$, calculated by Xavier Gourdon on October 27, 2000. (Caldwell, *Prime Pages*) The previous record was $\pi(10^{20}) = 2{,}220{,}819{,}602{,}560{,}918{,}840$, by Marc Deleglise and Paul Zimmermann. (Deleglise and Zimmermann 1996)

There are many estimates for the value of $\pi(n)$. For example, if $n > 16$, then $n/\log n < \pi(n) < n/(\log n - 3/2)$ (Williams 1998, 18)

$$\text{For } x > 1,\ \frac{n}{\log n} < \pi(n) < \frac{1.125506n}{\log n}$$

$$\text{For } x > 54,\ \frac{x}{\log x + 2} < \pi(x) < \frac{x}{\log x - 4}$$

The table and the following graph show how the values of $\pi(n)$ and several different estimates compare:

n	$\pi(n)$	$n/\log n$	$n/(\log n - 1)$	Gauss's Li	Legendre
10^3	168	145	169	178	172
10^4	1229	1086	1218	1246	1231
10^5	9592	8686	9512	9630	9588
10^6	78498	72382	78030	78628	78534
10^7	664579	620420	661459	664918	665138
10^8	5761455	5428681	5740304	5762209	5769341
10^9	50847478	48254942	50701542	50849235	50917519

In this table Gauss's Li(x) is always larger than $\pi(x)$. However, in 1914 Littlewood proved that $\pi(x) - \text{Li}(x)$ assumes both positive and negative values infinitely often. Skewes then proved in 1933, assuming the truth of the **Riemann hypothesis**, that the first switch occurs before x reaches $10^{10^{10^{34}}}$.

This was an extraordinarily large number at that time, "the largest number which has ever served any definite purpose in

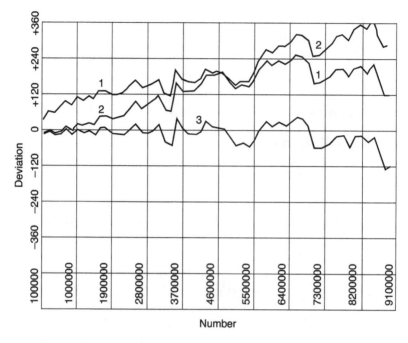

Note: 1 = Gauss   2 = Legendre   3 = Riemann

mathematics," according to G. H. Hardy, who compared it to the number of possible games of chess in which every particle in the universe was a piece.

Today, Skewes's number is negligible (!) compared to certain numbers appearing in conbinatorics, such as Graham's number, which can only be expressed using a special notation.

It is also a very poor estimate: in 1986 te Riele showed that between $6.62 \cdot 10^{370}$ and $6.69 \cdot 10^{370}$ there are more than 10^{180} consecutive integers for which Li(x) is an underestimate.

A much better approximation than any of these is the Riemann function, $R(x)$, which is defined using the Möbius function, $\mu(n)$, which is 1 for $n = 1$, 0 if n is *not* **squarefree**, and $(-1)^k$ otherwise, where k is the number of distinct prime factors in n. $R(x)$ is the sum from 1 to infinity of $\mu(n)/n$ Li($x^{1/n}$) =

$$\text{Li}(x) - \tfrac{1}{2}\,\text{Li}(x^{\frac{1}{2}}) - \tfrac{1}{3}\,\text{Li}(x^{\frac{1}{3}}) - \tfrac{1}{5}\,\text{Li}(x^{\frac{1}{5}}) + \tfrac{1}{6}\,\text{Li}(x^{\frac{1}{6}}) - \ldots$$

The differences between $\pi(x)$, Li(x), and $R(x)$ from 10^2 to 10^{16} are:

x	$\pi(x)$	$\text{Li}(x) - \pi(x)$	$R(x) - \pi(x)$
10^2	25	5	1
10^4	1,229	17	−2
10^6	78,498	130	29
10^8	5,761,455	754	97
10^{10}	455,052,511	3,104	−1,828
10^{12}	37,607,912,018	38,263	−1,476
10^{14}	3,204,941,750,802	314,890	−19,200
10^{16}	279,238,341,033,925	3,214,632	327,052

(Riesel 1994, 52)

In the 1870s Meissel developed a clever way to calculate $\pi(x)$ far beyond the known tables of primes and in 1885 (slightly mis-)calculated $\pi(10^9)$. Meissel's methods were simplified by D. H. Lehmer in 1959. Further improvements followed, and finally in October 2000 Xavier Gourdon and his **distributed computing** project calculated $\pi(10^{21})$.

estimating p(n)

The nth prime is denoted by $p(n)$ and is roughly $n \log n$. A recent estimate is that $0.91n \log n < p(n) < 1.7n \log n$. A better estimate for $p(n)$ is $n(\log n + \log \log n - 1)$, and Pierre Dusart has proved that this is always an underestimate. (Dusart 1999)

calculating p(n)

It is of course possible, with sufficient computing power, to calculate $p(n)$ exactly, though there is no very simple formula.

Here are some results:

The 1,000,000th prime	15,485,863
The 1,000,000,000th prime	22,801,763,489
The 10,000,000,000th prime	252,097,800,623
The 1,000,000,000,000,000th prime	37,124,508,045,065,437

a curiosity

Solomon Golomb, who coined the term *pentomino* in a talk to the Harvard Math Club in 1953 and later invented Golomb's Game, proved that as n increases, the ratio $n/\pi(n)$ takes in due course every integer value greater than 1. (Golomb 1962)

These are the first values of n at which the values 2 to 6 appear:

n	$\pi(n)$	$n/\pi(n)$
2	1	2
27	9	3
96	24	4
330	66	5
1080	180	6

prime pretender

The title *prime pretender to base b* is given to any composite number, n, that satisfies the equation $n^b \equiv n \pmod{b}$. This is a less restrictive condition that **Fermat's Little Theorem**, which says that if b and n are coprime, then $n^{p-1} \equiv 1 \pmod{p}$, and so there are more exceptions than the pseudoprimes that misleadingly satisfy Fermat's Little Theorem.

The title *primary pretender* for base b is given to the smallest prime pretender to base b. Since the smallest **Carmichael number**, 561, satisfies $n^{p-1} \equiv 1 \pmod{b}$ for all bases, and therefore, $n^b \equiv n \pmod{b}$, every primary pretender is ≤ 561.

Conway and colleagues showed that there are only 132 distinct primary pretenders, and the sequence starts

4, 4, 341, 6, 4, 4, 6, 6, 4, 4, 6, 10, 4, 4, 14, 6, 4, 4, 6, 6, 4, 4, 6, 22, 4, 4, 9, 6, 4, 4, 6, 6, 4, 4, 6, 9, 4, 4, 38, 6, 4, 4, 6, 6, 4, 4, 6, 46, 4, 4, 10, 6, 4, 4, 6, 6, 4, 4, 6, 15, 4, 4, 9, 6, 4, 4, 6, 6, 4, 4, 6, 9, 4, 4, 15, 6, 4, 4, 6, 6, 4, 4, 6, 21, 4, 4, 10, . . . (Sloane A000790)

This sequence is periodic, with this 122-digit period:

1956858433346007258724534003773627898201721382933760433673 4362294738647777395483196097971852999259921329236506842360 439300

This number is $(2 \cdot 3 \cdot 5 \cdot 7 \cdot 11 \ldots 59)(2 \cdot 3 \cdot 5 \cdot 7 \cdot 11 \cdot 13 \cdot 17 \cdot 19 \cdot 23)$ or $59\# \cdot 23\#$ where $p\#$ is **primorial** p.

In this product, $p(59) = 277$ is the largest possible prime factor, and $p(9) = 23$ is the largest possible *repeated* prime factor, of a composite number less than the Carmichael number 561. (Conway et al. 1997)

primitive prime factor

In a sequence of integers, a prime factor of one term is called primitive if it does not divide any previous term. For example, the first few terms of the Fibonacci sequence have the primitive prime factors entered below:

1	1	2	3	5	8	13	21	34	55
		2	3	5		13	7	17	11

See Bang's theorem; Lucas numbers

primitive roots

The powers of 3 (mod 5) are:

3^n	3	9	27	81	243	729	...
3^n (mod 5)	3	4	2	1	3	4	...

The residues repeat, in the sequence 3-4-2-1, which includes all the integers less than 5. Therefore, 3 is a primitive root of 5.

In general a primitive root, g, of a prime p is such that the residues of $g, g^2, g^3, g^4, \ldots, g^{p-1}$ (mod p) are all distinct, which means that they are a permutation of the numbers 1 to $p - 1$.

It isn't necessary to actually work out the powers of g: it is enough to reduce each power (mod p) and then multiply by g. For example, the first few powers of 5 (mod 13) are:

$$5 = 5 \pmod{13}$$
$$5^2 = 25 \equiv 12 \pmod{13}$$
$$5^3 \equiv 5 \cdot 12 = 60 \equiv 8 \pmod{13}$$
$$5^4 \equiv 5 \cdot 8 = 40 \equiv 1 \pmod{13}$$

Since $5^4 \equiv 1$ (mod 13), the sequence of residues now repeats, and 5 is *not* a primitive root of 13.

Any odd prime, p, has $\phi(\phi(p)) = \phi(p - 1)$ primitive roots. So 13 has $\phi(12) = 4$ primitive roots, which are 2, 6, 7, and 11.

Not only primes have primitive roots. A composite integer n has a primitive root if (and only if) it is 1, 2, or 4 or is of one of the forms p^a or $2p^a$ where p is an odd prime. In this case, it has $\phi(\phi(n))$ primitive roots.

We can express this another way: if $g^{\phi(m)}$ is the smallest power of g that is congruent to 1 (mod m), then g is a primitive root of m.

Johann H. Lambert (1728–1777) announced without proof that every prime number has at least *one* primitive root.

There is no fast algorithm for finding primitive roots of a general integer, or even for testing whether a number is a primitive root.

These are the smallest primitive roots of the primes in sequence:

prime	2	3	5	7	11	13	17	19	23	...
smallest primitive root	1	2	2	3	2	2	3	2	5	...

The next "records" are 6, which is the least primitive root of 41, and 7, which is the least primitive root of 71. (Sloane A001918)

These are the primes with 2 as a primitive root:

3   5   11   13   19   29   37   53   59   61   67   83   ...

(Sloane A001122)

Artin's conjecture

Artin conjectured in 1927 that if n is neither −1, 0, or 1 nor a perfect square, then the number of primes for which n is a primitive root is infinite. This question has not been settled, but it is known that it is true for $n = 2$, 3, and 5.

There is also a connection with recurring **decimal** periods. If p is a prime greater than 5, then the decimal expansion of $1/p$ has the maximum possible period of $p - 1$ if and only if 10 is a primitive root modulo p. This was the subject of another famous conjecture by Artin.

a curiosity

The smallest prime, p, such that its smallest primitive root is not a primitive root of p^2, is 40487. (Dr. Glasby: Caldwell, *Prime Pages*)
  See decimals, recurring

primorial

If p is prime, then primorial p, denoted by $p\#$, is the product of all the primes less than or equal to p. The sequence of primorials starts

1   2   6   30   210   2310   30030   510510   9699690   223092870
6469693230   200560490130 . . . (Sloane A002110)

primorial primes

Primorial primes are of the form $p\# \pm 1$. George Pólya is reputed to
have replied to a pupil who asked how often the product of the first
n primes, plus 1, was itself prime, "There are many questions which
fools can ask that wise men cannot answer." Gauss is supposed to
have made much the same answer to the question why did he not
tackle **Fermat's Last Theorem**. (Golomb 1981)

It is not known if $p\# + 1$ contains an infinity of primes, or even if an
infinity are composite. The five largest known primorial primes are,

primorial prime	no. of digits	year	discoverer
392113# + 1	169,966	2001	Daniel Heuer
366439# + 1	158,936	2001	Daniel Heuer
145823# + 1	63,142	2000	A. E. Anderson
			and D. E. Robinson
42209# + 1	18,241	1999	Chris Caldwell
24029# + 1	10,387	1993	Chris Caldwell
			(Caldwell, *Prime Pages*)

Let $Q_n = (p_1 p_2 p_3 \ldots p_n)^{1/p_n}$. Then the limit of Q_n as n tends to infinity is e, the base of natural logarithms. (Ruiz 1997)

Proth's theorem

François Proth (1852–1879) was a farmer and self-taught mathematician who died at the age of twenty-seven, having spent his entire life
in the village of Vaux-devant-Damloup near Verdun. (Williams 1998,
121) One of the many glories of mathematics is that progress can be
and has been made by the humble and the amateur. This is not to
say that the research "boundary" of mathematics is not usually far
distant from amateur studies—it is—but rather that mathematics is so
rich and so fertile, and there are so many ideas left to discover, that
the amateur on occasion can make a great contribution.

In 1878 he proved Proth's theorem: if $N = k \cdot 2^n + 1$ with $2^n > k$ and
if there is an integer a such that N divides $a^{(N-1)/2} + 1$, then N is
prime. This test is extremely simple and covers **Fermat numbers**,
Cullen numbers, and others.

One of the most popular prime number programs on the Internet is the Proth.exe program created by Yves Gallot in 1997, originally to find larger factors of the Fermat numbers that are known to be of the form $k \cdot 2^n + 1$. It now also handles numbers of the form $k \cdot 2^n - 1$, and **Generalized Fermat numbers** of the form $b^{2^n} + 1$. In 1997 Carlos Rivera used it to find the largest known "non-Mersenne" prime, $9183 \cdot 2^{262112} + 1$. In 1998, Chip Kerchner found the then largest known **Sophie Germain prime**, $92305 \cdot 2^{16998} + 1$.

pseudoperfect numbers

Sierpinski called numbers that are the sum of some of their factors only, *pseudoperfect*. For example, $12 = 2 + 4 + 6$. If we divide by 12, we get an example of Egyptian fractions, which sum to unity:

$$1 = \tfrac{1}{6} + \tfrac{1}{3} + \tfrac{1}{2}$$

Similarly, $20 = 10 + 5 + 4 + 1$ and $1 = \tfrac{1}{2} + \tfrac{1}{4} + \tfrac{1}{5} + \tfrac{1}{20}$.

Conversely, any expression of the form $1 = \tfrac{1}{a} + \tfrac{1}{b} + \tfrac{1}{c} + \ldots$ leads to a pseudoperfect number. This is an example of a set of reciprocals of odd integers that sum to 1:

$$1 = \tfrac{1}{3} + \tfrac{1}{5} + \tfrac{1}{7} + \tfrac{1}{9} + \tfrac{1}{15} + \tfrac{1}{21} + \tfrac{1}{27} + \tfrac{1}{35} + \tfrac{1}{63} + \tfrac{1}{105} + \tfrac{1}{135}$$

(Guy 1981, D11: 89)

It is equivalent to:

$945 = 315 + 189 + 135 + 105 + 63 + 45 + 35 + 27 + 15 + 9 + 7$

See weird numbers

pseudoprimes

The problem with the simple tests that a prime number must pass is that some composite numbers satisfy them also. In other words, passing the test is necessary but not sufficient to prove that the number is prime.

So, for every simple test there are some composite numbers that pass it. These are called *pseudoprimes*. The simplest pseudoprimes are the Fermat pseudoprimes, which satisfy **Fermat's Little Theorem**, although they are actually composite. (If the term *pseudoprime* is used without qualification a Fermat pseudoprime is meant.)

According to Fermat's Little Theorem, if p is prime and a and p are coprime, then

$$a^{p-1} \equiv 1 \pmod{p}$$

Unfortunately the converse is not true. Sometimes,

$$a^{n-1} \equiv 1 \pmod{n}$$

although n is composite. For example,

$$2^{341-1} \equiv 1 \pmod{341}$$

although $341 = 11 \cdot 31$.

This is the smallest exception with $a = 2$ and is therefore the smallest Fermat pseudoprime to base 2, discovered by Sarus in 1819. (Base 2 pseudoprimes are occasionally called Sarus numbers; also Poulet numbers, after Paul Poulet, who published an early table of pseudoprimes in 1938; and occasionally Fermatians.)

There are also pseudoprimes to other bases: the smallest to base 3 is $91 = 7 \cdot 13$, and to base 5, $217 = 7 \cdot 31$. (If you count *even* pseudoprimes, then $124 = 2^2 \cdot 31$ is the smallest in base 5. Some authors simply don't count even numbers as pseudoprimes, since it is so obvious that they are composite anyway.)

We could have discovered that 341 is composite by checking it using another base, for example, 3:

$$3^{341-1} \equiv 56 \pmod{341}$$

So 341 is composite after all. It is a pseudoprime to base 2 but not to base 3. By checking to see if a number is pseudoprime to various different bases, we can usually discover quite quickly if it is composite. However, some rare numbers are pseudoprimes to *every* base. These are the **Carmichael numbers**.

The sequence of pseudoprimes to base 2 starts: 341, 561, 645, 1105, 1387, 1729, 1905, 2047, . . . (Sloane A001567) These are all odd. D. H. Lehmer found the smallest even pseudoprime to base 2 which is $161038 = 2 \cdot 73 \cdot 1103$, since $2^{161038} \equiv 2 \pmod{161038}$.

There is an infinity of pseudoprimes to any given base, and an infinity of pseudoprimes with exactly k factors, provided $k > 1$. However, pseudoprimes are much scarcer than genuine primes. There are 78,498 primes less than 1,000,000 but only 245 pseudoprimes to base 2.

bases and pseudoprimes

To which bases is 91 a pseudoprime? It is a pseudoprime to thirty-six bases, starting with: 1, 3, 4, 9, 10, 12, 16, 17, 22, 23, 25, 27, 29, 30, 36, 38, 40, 43, 48, 51, ... It is no coincidence that $\phi(91) = 72 = 2 \cdot 36$. The ratio of $\phi(n)$ to the number of bases to which it is a pseudoprime is 1 for prime numbers and **Carmichael numbers**, and 2, 3, 4, or 5 for these numbers:

2: 4, 6, 15, **91**, 703, 1891, 2701, 11305, 12403, 13981, 18721, 23001, . . .
3: 9, 21, 45, 65, 105, 133, 231, 341, 481, 645, 1541, 3201, 4033, 4371, . . .
4: 8, 10, 12, 28, 66, 85, 435, 451, 946, 1387, 2047, 3277, 3367, 5551, . . .
5: 25, 33, 165, 217, 325, 385, 793, 1045, 1065, 2665, 3565, 4123, . . .

(Gérard Michon, http://home.att.net/~numericana)

pseudoprimes, strong

Starting with an odd number, n, and a base b, we know that if,

$$b^{n-1} \equiv 1 \pmod{n}$$

then either n is prime or it is pseudoprime to base b. Now write n in the form $2^k \cdot m + 1$. Then,

$$b^{2^k m} \equiv 1 \pmod{n}$$
$$b^{2^k m} - 1 \equiv 0 \pmod{n}$$

So, by factorizing $b^{2^k m} - 1$

$$(b^m - 1)(b^m + 1)(b^{2m} + 1)(b^{4m} + 1) \ldots (b^{2^{(k-1)}m} + 1) \equiv 0 \pmod{n}$$

Therefore, n divides one of these factors, if it is prime. If it is composite but divides one of these factors anyway, then it is a *strong pseudoprime to base b*.

Testing whether n does indeed divide any of these expressions is a stronger test than the usual pseudoprime test, because strong pseudoprimes are extremely rare.

There are just three strong pseudoprimes less than 1,000, and only forty-six below 1,000,000, compared to 245 ordinary pseudoprimes to base 2. Moreover, there are no **Carmichael numbers** among the strong pseudoprimes. If n is composite, then it will fail the strong pseudoprime test for at least half the bases less than n.

public key encryption

Substitution ciphers have been found in Egyptian hieroglyphics and in Mesopotamian cuneiform tablets. They were later used by Julius Caesar, who gave his name to the Caesar Cipher. Leaping forward across two thousand years, the Second World War was shortened by the success of the mathematicians at Bletchley Park, including Alan Turing, who cracked the codes produced by the German Enigma machine, and where Colossus, the world's first electronic computer, was used from 1944 onward. More recently, the British government has announced the creation of a new institute devoted to "signals intelligence," or the study of communications "used by terrorists and criminals."

Traditional methods of sending secret mesages, although ingenious, shared a problem, which was that the sender and the recipient needed the same key, and exchanging the key was a risky business because it could too easily be intercepted.

Whitfield Diffie and Martin Hellman in 1976 had a bright idea: instead of having the same key for encryption and decryption, what about having *different* keys? Then the encryption key could be openly published, but anyone with the coded message in their hands would be helpless to decrypt the message. There would, of course, have to be some connection between the two keys, but this ought to be as obscure as possible, except to the decoder. Alternatively, you might publish the decryption key but hide the encryption method. (It was later revealed that the British Secret Service had already thought of the same idea but—force of habit?—kept it secret!)

The original idea that Diffie and Hellman published in the prophetically titled "New Directions in Cryptography" went like this. First, two numbers are made public, a *prime p*, and a *generator g*, which is chosen so that every number from 1 to $p - 1$ is of the form g^n for some value of n.

Alan and Betty can then create shared secret keys as follows: Alan chooses his own random number a and calculates g^a (mod p), while Betty chooses her number b and calculates g^b (mod p). They then exchange the results. Alan then calculates $(g^b)^a$ (mod p) and Betty calculates $(g^a)^b$ (mod p). They now share the secret key g^{ab} (mod p), and this will remain secret as long as it is impossible, or simply impracticable, to calculate g^{ab} from g^a and g^b, should the latter become public knowledge because they exchanged them over an insecure channel. This is so provided the prime p is large enough.

The result was a range of public key encryption systems, most of which turned out not to be as unbreakable as their inventors fondly imagined: the clear winner in the Darwinian struggle was the **RSA algorithm**.

These are not used in practice to encrypt entire messages, because they are about a thousand times slower than conventional cryptography, but they can be used to encrypt the conventional cryptographic key.

See RSA algorithm

pyramid, prime

Margaret Kenney suggests arranging the numbers 1 to n in each row of a pyramid, so that each adjacent pair of numbers sums to a prime:

```
            1       2
         1     2       3
      1     2     3       4
   1     4     3     2       5
1     4     3     2     5       6
```

and so on. Every row begins with 1, and ends with n. (www .geometry.net, Pascal's Triangle Geometry)

There are now two possibilities for row 7:

```
     1  4  3  2  5  6  7
or   1  6  5  2  3  4  7
```

(Weisstein, *MathWorld*)

The number of ways of completing each row is Sloane sequence number A036440:

$$0, 1, 1, 1, 1, 1, 2, 4, 7, 24, 80, \ldots$$

See circle, prime

Pythagorean triangles, prime

Pythagorean triangles are right-angled triangles with integral sides. The lengths of the sides can always be represented in the form, $2mn$, $m^2 - n^2$, $m^2 + n^2$, for some values of m and n.

If **hypothesis H** of Sierpinski and Schinzel is true, then there is an infinity of Pythagorean triangles with one leg and the hypotenuse both prime. Harvey Dubner and Tony Forbes have searched for such triangles. The largest they found had one leg of 5,357 digits and a hypotenuse of 10,713 digits. (Dubner and Forbes 2001)

quadratic residues

If we take a prime, p, and square all the integers less than p, we get a limited number of answers. For example, if $p = 11$, then

x^2	1^2	2^2	3^2	4^2	5^2	6^2	7^2	8^2	9^2	10^2
x^2 (mod 11)	1	4	9	5	3	3	5	9	4	1

The pattern is (inevitably) symmetrical. The numbers 1, 3, 4, 5, and 9 are the quadratic residues of 11, or the quadratic residues modulo 11. The numbers 2, 6, 7, and 8 are the non-residues.

If p is an odd prime, it has $(p - 1)/2$ quadratic residues and the same number of non-residues.

The number -1 is a quadratic residue of primes of the form $4n + 1$ and a non-residue of primes $4n + 3$. It follows that if the prime $p = 4n + 1$, then there is a multiple of p that is of the form $1 + x^2$. Similarly, 2 is a quadratic residue of primes of the form $8n \pm 1$ and a non-residue of primes $8n \pm 3$.

Euler's criterion says that a is a quadratic residue of p if $a^{(p-1)/2} \equiv 1$ (mod p) and a non-residue if $a^{(p-1)/2} \equiv -1$ mod p. Notice that it must be one or the other! So the 5th powers of 1, 3, 4, 5, and 9 $\equiv$ 1 (mod 11), while 2, 6, 7, 8, and 10 $\equiv$ -1 (mod 11).

The product of two residues or two non-residues is a residue, but the product of a residue and a non-residue is a non-residue—they behave like positive and negative numbers.

If $a \equiv b$ (mod p), then a and b are either both residues or both non-residues.

residual curiosities

If N is the maximum number of consecutive non-residues of a prime p, Hudson proved the conjecture that $N < \sqrt{p}$, if p is large enough. The only exception seems to be $p = 13$. (Hudson: Guy 1994)

Issia Schur has conjectured that 13 is the only prime, p, with more than $\sqrt{p}$ consecutive non-residues. Hudson has proved this is so for $p > 2^{232}$. (Caldwell, *Prime Pages*)

Quadratic residues often appear in applications of mathematics, for example in the design of the best reception in concert halls, and in noise abatement structures. Concert hall acoustics also create problems of normal modes of vibration in cubical or near-cubical resonators, whose solution depends on representing integers as the sum of three squares. (Schroeder 1992)

polynomial congruences

What about more complicated polynomials than the simple x^2? Lagrange (1736–1813) proved that if p is a prime, then the polynomial congruence,

$$a_n x^n + a_{n-1} x^{n-1} + \ldots + a_1 x + a_0 \equiv 0 \text{ (mod } p)$$

has, *at most*, n solutions (mod p), provided a_n is not divisible by p.

For example, $x^3 - 5x + 3 \equiv 0$ (mod 7) has the single solution $x \equiv 6$ (mod 7), while $x^3 - 4x + 9$ has three solutions (mod 3), which means that all its values are divisible by 3.

See Gaussian primes; quadratic reciprocity, law of

quadratic reciprocity, law of

This law was discovered by Euler in 1772, partly proved by Legendre in 1785, and rediscovered by Gauss, at the age of eighteen, who proved it in total of eight different ways, the first in 1796, and called it the *aureum theorema*, or golden theorem.

Frank Lemmermeyer published in 2000 a list of 196 different proofs, but this is already out of date. At www.rzuser.uni-heidelberg.de there is a list of 207 proofs, 8 of them in the new millennium, evidence of the importance, depth, and fascination of this theorem.

It can be most briefly expressed using the Legendre symbol:

$(a|q)$ equals: $+1$ if a is a quadratic residue of p
                   -1 if a is not a quadratic residue of p

and $(n|p)$ equals: 0 if $n \equiv 0 \pmod{p}$

Legendre's function might seem bizarre, but it is actually very useful, not least because it has several simple properties. For example, if p is an odd prime, then for any two integers a and a',

$$(aa'|p) = (a|p)(a'|p)$$
$$(a^2|p) = 1$$

and if $a \equiv b \pmod{p}$ then $(a|p) = (b|p)$

- The first property implies that $(a^r|p) = (a|p)^r$.
- The first two properties imply that $(a^2 b|p) = (b|p)$.

The law of quadratic reciprocity states that if p and q are distinct odd primes, then $(q|p) = (p|q)$ unless both p and q are of the form $4n + 3$, in which case $(q|p) = -(p|q)$.

Notice that, once again, a subtle but striking difference appears between numbers of the two forms $4n + 1$ and $4n + 3$!

The law can also be expressed as $(p|q) = (-1)^{(p-1)(q-1)/4}(q|p)$.

It can now be used to calculate values of $(a|b)$. For example, is 152 a quadratic residue modulo 43? Let's see:

$$(152|43) = (23|43) = -(43|23) = -(20|23)$$
$$= -(2^2|23)(5|23) = -(1)(5|23) = -(23|5) = -(3|5)$$
$$= -(5|3) = -(2|3) = -(-1) = 1$$

Or, alternatively:

$$(152|43) = (2^3|43)(19|43) = (2|43)^3(19|43)$$
$$= (-1)^3(-1)(43|19) = (5|19) = (19|5) = (4|5) = (2^2|5) = 1$$

Euler's criterion

If p is an odd prime, then for all n, $(n|p) = n^{(p-1)/2} \pmod{p}$.

It follows that: $(-1|p) = 1$ if $p \equiv 1 \pmod 4$
$(-1|p) = -1$ if $p \equiv 3 \pmod 4$

Similarly, $(2|p) = 1$ if $p \equiv \pm 1 \pmod 8$
$(2|p) = -1$ if $p \equiv \pm 3 \pmod 8$

See Hilbert's 23 problems; quadratic residues

Ramanujan, Srinivasa (1887–1920)

An equation for me has no meaning unless it expresses a thought of God.

—*Ramanujan (Hoffman 1998, 85)*

Srinivasa Aiyangar Ramanujan was born in Erode, a small town in Tamil Nadu state, India. He was a self-taught mathematician with an uncanny ability to manipulate formulae reminiscent of Euler. Suffering from poverty and moving from one petty job to another as he pursued his mathematical inspirations, he eventually wrote to one mathematician in England—no response—then another—no response—and then finally to G. H. Hardy, who recognized his genius, explaining that his theorems "must be true, because, if they were not true, no one would have had the imagination to invent them." (Hardy 1940, 9) Hardy arranged for him to come to Cambridge, persuading Madras University to grant him a research scholarship of £250 a year for five years plus £100 for his passage to England in 1914.

Hardy had a very high opinion of Ramanujan, at least in terms of pure "natural" talent, for which he rated him at 100, giving himself 25, his longtime collaborator Littlewood 30, and the great German, David Hilbert, 80. (Hardy's judgment of himself is suspect, not to say bizarre: he had a pessimistic streak in his makeup, which came out in his famous work, *A Mathematician's Apology*, which was indeed "apologetic," as if his wonderful life's work needed any excuse. On one occasion he tried to kill himself, as did indeed Ramanujan. Fortunately, they both failed.)

Ramanujan produced, with Hardy, some remarkable mathematics, and left behind him his extraordinary notebooks, which have since been transcribed and published and continue to be studied, but the English climate and food did not agree with him. He contracted tuberculosis, and returned to India in 1918, to die two years later.

Ramanujan had his own view of his talents: "Ramanujan and his family were ardent devotees of God Narasimha (the lion-faced incarnation of God), the sign of whose grace consisted in drops of blood seen during dreams. Ramanujan stated that after seeing such drops, scrolls containing the most complicated mathematics used to unfold before him and that after waking, he could set down on paper only a fraction of what was shown to him." (Ranganathan 1967, 88)

Ramanujan had a truly wonderful gift for manipulating formulae of all kinds. For example, from the fact that,

$$1/2^2 + 1/3^2 + 1/4^2 + 1/5^2 + \ldots = \pi^2/6$$

he deduced that,

$$1/2^2 + 1/3^2 + 1/5^2 + 1/7^2 + 1/8^2 + 1/11^2 + 1/12^2 + \ldots = \pi^2/20$$

where the denominators are the integers with an *odd* number of prime divisors. Similarly,

$$1/2^2 + 1/3^2 + 1/5^2 + 1/7^2 + 1/11^2 + 1/13^2 + 1/17^2 + \ldots = 9/2\pi^2$$

where the denominators are the integers containing an *odd* number of *dissimilar* prime divisors. (Ramanujan 1913)

highly composite numbers

One of Ramanujan's longest and best papers was on highly composite numbers, which he defined as a number whose "number of divisors exceeds that of all its predecessors." He proved that if $2^a \cdot 3^b \cdot 5^c \cdot 7^d \cdot \ldots p^z$ is highly composite, then $a \geq b \geq c \geq d \ldots \geq z$ and that $z = 1$, for all highly composite numbers except 4 and 36. (Ramanujan 1915)

He also proved that the ratio of very large consecutive highly composite numbers tends to 1 and gave a list of 103 highly composite numbers, which starts:

2, 4, 6, 12, 24, 36, 48, 60, 120, 180, 240, 360, 720, 840, . . .

There are just six highly composite numbers, if 1 is included, that are divisors of every larger highly composite number: (1), 2, 6, 12, 60, and 2520. (Steven Ratering 1991)

It is possible to find a sequence of consecutive integers, as long as we choose, whose members are all highly composite. (Subrakamian and Becker 1966)

randomness, of primes

To put it poetically, primes play a game of chance.

—Mark Kac (1959)

It is evident that the primes are randomly distributed but, unfortunately we don't know what "random" means.

—R. C. Vaughan (Chance News 11.02)

The prime numbers are so irregular that is tempting to think of them as some kind of random sequence, in which case it should be possible to use the theory of probability and statistics to study them. The first and most famous application of probability to primes was the Erdös-Kac theorem.

In 1917 **Hardy** and **Ramanujan** had proved that provided N is large, the "typical" number of prime factors is log log N. (A very small proportion of numbers has many small prime factors, but such numbers are untypical.) The *normal law* states, very roughly, that many distributions in nature behave as if they were the result of tossing a coin many times. Poincare claimed that there is something mysterious about the normal law because mathematicians think it is a law of nature but physicists believe it is a mathematical theorem. (Kac 1959)

Mark Kac, a pioneer in probability theory, fortunately collaborated with Paul **Erdös**, who was an expert in analytic number theory, to prove that it makes sense to think of the number of distinct prime factors of *N*, *if* N *is large*, as the sum of statistically independent functions $N(p)$, which are 1 if $p \mid N$ and 0 if $p \nmid N$. It follows that the number of prime factors of *N* fits the normal law, with a mean of log log *N* and a standard deviation of $\sqrt{(\log \log n)}$. (Erdös and Kac 1945)

More precisely, Kac and Erdös proved that the proportion of integers for which

$$\log \log n + a\sqrt{2 \log \log n} < d(n) < \log \log n + b\sqrt{2 \log \log n}$$

equals the area under the "normal" curve,

$$\frac{1}{\sqrt{\pi}} e^{-x^2} \text{ between } x = a \text{ and } x = b.$$

So, for example, the average number of prime factors is five round about 10^{70}, and more precisely, about 60% of the integers round about 10^{70} have between three and seven prime factors. Similar results can be proved for other functions, such as Euler's $\phi(n)$ or totient function.

Three physicists at Boston University, Pradeep Kumar, Plamen Ivanov, and Eugene Stanley, have been studying heartbeat rhythms. Now they believe they may have found a curious kind of order in the sequence of prime numbers. They studied the second differences in the sequence of primes:

	2	3	5	7	11	13	17	19	23	29	31
1st diffs.		1	2	2	4	2	4	2	4	6	2
2nd diffs.			+1	+0	+2	−2	+2	−2	+2	+2	−4	+4

If the primes really jump around randomly, then these second differences should jump randomly also, but Kumar and his colleagues concluded that they don't. Positive values are usually followed by (corresponding) negative values, for example, the first +2, −2 differences. They found an oscillation of period 3.

They also decided that second differences that are multiples of 6 are less frequent than other differences. This fits previous statistical analyses that found oscillations with period 6 in the distances between consecutive primes. (Kumar, Ivanov, and Stanley 2003)

Patrick Billingsley has explained how we can also think of a single composite number as generating a random walk. You take the

---

Kac and Erdös Collaborate

"Erdös, who was spending the year at the Institute of Advanced Study, was in the audience but he half-dozed through most of my lecture; the subject matter was too far removed from his interests. Toward the end I described briefly my difficulties with the number of prime divisors. At the mention of number theory Erdös perked up and asked me to explain once again what the difficulty was. Within the next few minutes, even before the lecture was over, he interrupted to announce that he had the solution!

"The reader, I hope, will forgive my lack of modesty if I say that it is a beautiful theorem. It marked the entry of the normal law, hitherto the property of gamblers, statisticians and *observateurs*, into number theory and . . . it gave birth to a new branch of this ancient discipline." (Kac 1987, 90–91)

---

primes, 2, 3, 5, . . . and move up one step if the prime is a factor of the number and down one step if it is not. The result is rather like repeated coin-tossing or Brownian motion—and by adjusting the size of the steps taken can be made more so. It is then possible to deduce theorems about prime numbers from well-known results in probability theory. (Billingsley 1973)

Von Sternach and a prime random walk

The reciprocal of the **Riemann zeta function**, $1/\zeta_{(s)}$, can be written as this series,

$$1/\zeta(s) = 1 - 1/2^s - 1/3^s + 0/4^s - 1/5^s + 1/6^s - 1/7^s + 0/8^s + 0/9^s + 1/10^s \ldots$$

The numerators are the values of the Möbius function, $\mu(n)$, which is zero if n is *not* **squarefree**, 1 if it has an even number of prime factors, and −1 if it has an odd number of prime factors.

Von Sternach in 1896 realized that the values of the Möbius function,

+1, −1, −1, 0, −1, +1, −1, 0, 0, +1, −1, 0, −1, +1, +1, 0, −1, 0, −1 . . .

could be thought of as a random walk in which a drunk either stands where he is (0), or staggers forward one step (+1), or stumbles back one step (−1). So he listed the first 150,000 values of $\mu(n)$ and calculated that the probability that it was non-zero was, apparently, $6/\pi^2$, or roughly 0.608 with +1 and −1 appearing as values more or less equally often.

The **Riemann hypothesis** is now equivalent to saying (roughly and crudely!) that our random walker does not stagger too far from the origin. (Vaughan, Chance News 11.02)

record primes

A small community of mathematicians is devoted to finding larger, higher, bigger prime number records. What's the point? Well, competition is one answer, competition that is fun, that tests even modern powerful computers to their limits, and that has even become a sociable activity with the arrival of **distributed computing** in which the participants in projects such as **GIMPS, the Great Internet Mersenne Prime Search**, work in teams.

Finding the largest primes also depends not only on more powerful hardware but on more and more efficient algorithms that are beautiful in themselves: what could be more astonishing than proving beyond doubt that a number so large that it could not be written out by hand in less than millions of years is in fact prime?

some records

The largest known prime was found by GIMPS member Josh Findley, who used a 2.4 GHz Pentium 4 Windows XP PC running for fourteen days to finally prove the number prime, on May 15, 2004. It is $2^{24036583} - 1$ and has 7,235,733 decimal digits, nearly a million more than the previous record. GIMPS project leaders now see the first ten-million-digit prime within reach, for which the **Electronic Frontier Foundation** is offering a $100,000 award.

The two previous records were:

- Michael Schafer, a chemical engineering student at Michigan State University, and GIMPS: $2^{20996011} - 1$ (6,320,430 digits), November 17, 2003.
- Michael Cameron, $2^{13466917} - 1$ (4,053,946 digits), November 14, 2001.

A Point of Priority

"Hurwitz knew about M_{4423} seconds before M_{4253} (because of the way the output was stacked). John Selfridge asked 'Does a machine result need to be observed by a human before it can be said to be "discovered"?' To which Hurwitz replied, 'forgetting about whether the computer knew, what if the computer operator who piled up the output looked?' I [have] decided that Hurwitz discovered the prime when he read the output, so M_{4253} was never the largest known prime." (Caldwell, *Prime Pages*)

The non-Mersenne prime record is by Jeffrey Young: $3 \cdot 2^{303093} + 1$ (91,241 digits), found in 1998.

The record for a prime that is not of a special and easily tested form is $10^{9999} + 33603$ (10,000 digits), found using ECPP on August 19, 2003, by Jens Franke, Thorsten Kleinjung, and Tobias Wirth with a distributed version of their own ECPP program (www.ellipsa.net/primo/record).

Chris Caldwell speculates, based on the dates of previous record-breaking primes, that a 10,000,000-digit prime will be found by 2005, a 100,000,000-digit prime by 2015, and a 1,000,000,000-digit prime by early 2025. (Caldwell, *Prime Pages*)

See also aliquot sequences; arithmetic progressions, of primes; Brier numbers; Cullen primes; Cunningham chains; factorial primes; Fermat's Last Theorem; GIMPS; k-tuples conjecture, prime; Mersenne numbers; perfect, multiply; primorial; repunits, prime; Sophie Germain primes; twin primes; unitary divisors; Woodall primes

repunits, prime

The repunits, in base 10, are the numbers whose digits are all 1s, such as $R_5 = 11111$. Since $R_5 = 99999/9 = (10^5 - 1)/9$, the primality of R_n depends on the factorization of $10^n - 1$.

A repunit R_n can only be a prime if n itself is a prime. It is believed that there are infinitely many repunit primes, but only five are known: R_2, R_{19}, R_{23}, and R_{317}, which was proved prime by J. Brillhart and H. C. Williams in 1977, and R_{1031}, proved prime by Williams and Dubner in 1986.

The two largest known candidate repunit primes are both *proba-ble* primes: $R_{49081} = (10^{49081} - 1)/9$ was found by Harvey Dubner on September 9, 1999, and R_{86453} was discovered by Lew Baxter in October 2000.

In base a, the repunits are of the form $(a^n - 1)/(a - 1)$. A repunit in base 10 will therefore not be a repunit in another base unless we can solve the equation,

$$111 \ldots 1_{10} = 1 + a + a^2 + \ldots + a^n$$

This has no known solution. However, the equation,

$$1 + 5 + 5^2 = 31 = 1 + 2 + 2^2 + 2^3 + 2^4$$

means that $(5^3 - 1)(5 - 1) = (2^5 - 1)(2 - 1)$, the only known example of equal repunits to different bases.

In base 3, the repunits for $n \leq 1000$ are prime for $n = 3, 7, 13, 71, 103$, and 541.

In base 5, the repunits for $n \leq 1000$ are prime for $n = 3, 7, 11, 13, 47$, 127, 149, 181, 619, and 929. (Ribenboim 1988, 279)

A Curiosity

The number $140800 = 2^9 \cdot 5^2 \cdot 11$, so its prime factors sum to 39. When written to these bases,

base	1st digit	2nd digit	3rd digit
198	3	117	22
832		169	192
1200		117	400
1540		91	660
1728		81	832
2024		69	1144
2360		59	1560
2720		51	2080

in every case, the product of the digits when divided by the base equals 39. (www.mathpages.com/home/kmath083.htm)

Rhonda numbers

The sum of the prime factors of $25662 = 2 \cdot 3 \cdot 7 \cdot 13 \cdot 47$ is 72, and the product of the digits is 720, which is 10 times 72, so 25662 is a Rhonda number: the product of the digits in base b is equal to b times the sum of the prime factors.

There is an infinity of Rhonda numbers, including thirty-eight in base 10 below 200,000. (www.mathpages.com/home/math007.htm, "Smith Numbers and Rhonda Numbers")
  See Smith numbers

Riemann hypothesis

His style was conceptual rather than algorithmic—and to a higher degree than that if any mathematician before him. He never tried to conceal his thoughts in a thicket of formulas. After more than a century his papers are still so modern that any mathematician can read them without historical comment, and with intense pleasure.

—*Freudenthal (1975)*

Georg Friedrich Bernhard Riemann (1826–1866) was another prodigy. When still a youth, he was lent a copy of Legendre's *Treatise on the Theory of Numbers* and read the 900 pages in six days. He studied under Dirichlet, but presented to Gauss his thesis for a higher degree, *On the Hypotheses Which Lie at the Bases of Geometry*. It was read on June 10, 1854, less than a year before Gauss's death.

In 1862 Riemann married Elise Koch, but in the same year he caught a heavy cold and was then diagnosed with tuberculosis. His health had never been robust, and despite traveling to Italy in search of a cure, he died of the disease at the age of thirty-nine, leaving behind a few papers of the very highest quality.

In 1859 Riemann presented to the Berlin Academy the only paper he ever wrote on number theory, "On the Number of Prime Numbers Less Than a Given Quantity." This eight-page paper, obscure, condensed, and lacking in rigor by modern standards, contained a number of conjectures, none of which were proved in the next thirty years, but which provided ideas and problems for many of the greatest mathematicians who followed him.

In particular he discussed what is now called the Riemann zeta function:

$$\zeta(s) = 1/1^s + 1/2^s + 1/3^s + 1/4^s + 1/5^s + \ldots$$

which is the same as the series that Euler considered (page 72), except that for Riemann, s was a complex number, whereas for Euler it was real.

The Riemann zeta function has what are called the *trivial zeros* at $-2, -4, -6, \ldots$ The remaining zeros are all in the strip for which the real part of s lies between 0 and 1 inclusive and they are symmetrical about the line real $s = \frac{1}{2}$.

Riemann conjectured that this function has an infinite number of zeros with real part between 0 and 1, inclusive, and gave a formula, also conjectural, for the number of zeros of the function, *and then remarked that all the non-trivial zeros have real part ½.* This is the famous Riemann hypothesis—Riemann thought it "very likely"— which is universally regarded as the most important unsolved problem in mathematics.

The study of the distribution of the zeros of $\zeta(s)$ is important because the size of the error term in the **prime number theorem** depends on it. The more that is known about the zero-free region of $\zeta(s)$, the smaller the error term. The Riemann hypothesis is equivalent to $\pi(x) = \text{Li}(x)$ plus an error term that is bounded as x tends to infinity by a function $x^{1/2+e}$.

J. P. Gram in 1903 published a list of the first ten roots plus five larger roots. The first ten roots all have real part ½, and are approximately $\frac{1}{2} + ia$, where a has the values:

$$14.135 \quad 21.022 \quad 25.011 \quad 30.425 \quad 32.935$$
$$37.586 \quad 40.919 \quad 43.327 \quad 48.005 \quad 49.774$$

(Edwards 1974, 96)

G. H. Hardy proved in 1915 that there is an infinity of zeros on the critical line, and Conrey in 1989 that more than 40% of the zeros are on the line.

ZetaGrid

"ZetaGrid is a platform independent grid system that uses idle CPU cycles from participating computers. Grid computing can be used for any CPU intensive application which can be split into many separate steps and which would require very long computation times on a single computer. ZetaGrid can be run as a low-priority background process on various platforms like Windows, Linux, AIX, Solaris, and Mac OSX. On Windows systems it may also be run in screen saver mode. . . .

"At the IBM Development Laboratory in Böblingen ZetaGrid solves one problem in practice, running on six different platforms: The verification of Riemann's Hypothesis is considered to be one of modern mathematics' most important problems.

"This implementation involves more than 11,000 workstations and has a peak performance rate of about 7056 GFLOPS. More than 1 billion zeros for the zeta function are calculated every day." (www.zetagrid.net/zeta)

Bohr and Landau (1914) proved that for any $e > 0$, all but an infinitesimal proportion of roots lies within e of the line real $s = \frac{1}{2}$.

Rosser, Yohe, and Schoenfeld have calculated up to the $3\frac{1}{2}$ millionth zero—and they are all on real $s = \frac{1}{2}$. The calculations, which are assumed to be error-free, are on three reels of magnetic tape stored at the Mathematics Research Center in Madison, Wisconsin.

More recently, A. M. Odlyzko and A. Schönhage used a faster method of calculation to calculate 10^6 roots near root number 10^{20} and 10 billion roots near root number 10^{22}. (Odlyzko and Schönhage 1988)

One way to follow the behavior of $\zeta(s)$ is to study a different function, $Z(t) = e^{i\theta(t)}\zeta(\frac{1}{2} + it)$, which has real values but also has the same zeros as $\zeta(s)$. (The meaning of $e^{i\theta(t)}$ we shall not explain.)

The values of $Z(t)$ can be calculated using the Riemann-Siegel formula that was published by Siegel after he had made a careful study of Riemann's unpublished notes.

What does $Z(t)$ look like? Since it is intimately connected to $\zeta(s)$, which is intimately connected to the prime numbers, we might expect it to be very irregular, and, sure enough, it is! In particular, it

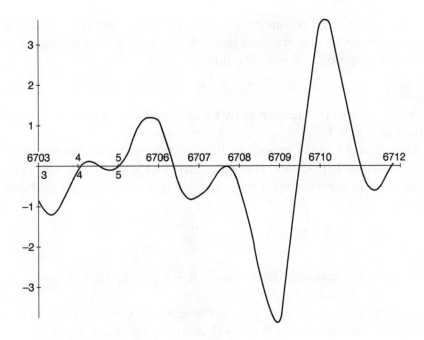

occasionally *almost fails to cross the x-axis,* which is extremely significant because a maximum just below the axis or a minimum just above would mean that *Riemann's hypothesis is false!* These events are referred to as Lehmer's phenomenon and are illustrated in the figure on this page. (Lehmer 1956)

How the gods are laughing at us! The point where the graph seems to touch the axis is actually a pair of zeros that is extremely close together: just as we have caught our breath, $Z(t)$ executes an enormous swing down and then up again! No wonder that Harold Edwards remarked, "The degree of irregularity of Z shown by this graph of Lehmer, and especially the low maximum value [illustrated] must give pause to even the most convinced believer in the Riemann hypothesis." (Edwards 1974, 178) (If you want to observe the behavior of $Z(t)$ yourself, go to www.math.ubc.ca/~pugh/RiemannZeta/ RiemannZetaLong.html, where there is a Java applet that plots $Z(t)$ in real time.)

the Farey sequence and the Riemann hypothesis

John Farey (1766–1826) published about sixty scientific articles and just one on mathematics, "On a curious property of vulgar fractions" (*Philosophical Magazine*, 1816), in which he pointed out the "curious property" that no one seemed previously to have spotted, that if you arrange all the fractions with denominators (for example) at most five, in order of size, like this,

$$\tfrac{1}{5} \quad \tfrac{1}{4} \quad \tfrac{1}{3} \quad \tfrac{2}{5} \quad \tfrac{1}{2} \quad \tfrac{3}{5} \quad \tfrac{2}{3} \quad \tfrac{3}{4} \quad \tfrac{4}{5} \quad 1$$

then they have the neat property that if $\tfrac{p}{q}$ and $\tfrac{r}{s}$ are successive terms, then $qr - ps = 1$.

Franel and Landau proved in 1924 that there is a connection with the Riemann hypothesis. There are ten terms in this Farey sequence that we can compare with the sequence of equally spaced fractions from $\tfrac{1}{10}$ to $\tfrac{10}{10}$:

$\tfrac{1}{5}$	$\tfrac{1}{4}$	$\tfrac{1}{3}$	$\tfrac{2}{5}$	$\tfrac{1}{2}$	$\tfrac{3}{5}$	$\tfrac{2}{3}$	$\tfrac{3}{4}$	$\tfrac{4}{5}$	1
$\tfrac{1}{10}$	$\tfrac{2}{10}$	$\tfrac{3}{10}$	$\tfrac{4}{10}$	$\tfrac{5}{10}$	$\tfrac{6}{10}$	$\tfrac{7}{10}$	$\tfrac{8}{10}$	$\tfrac{9}{10}$	$\tfrac{10}{10}$

Calculate the absolute differences: $|\tfrac{1}{5} - \tfrac{1}{10}| = \tfrac{1}{10}$; $|\tfrac{1}{4} - \tfrac{2}{10}| = \tfrac{1}{20} \ldots$ $|\tfrac{4}{5} - \tfrac{9}{10}| = \tfrac{1}{10}$ and $|1 - \tfrac{10}{10}| = 0$.

Now add up these absolute differences and call the sum S. Then the Riemann hypothesis is equivalent to the statement that $S/x^{(1/2+e)}$ tends to zero as x tends to infinity whatever the value of e. (Edwards 1974, 264)

the Riemann hypothesis and $\sigma(n)$, the sum of divisors function

Let $H_n = 1 + \tfrac{1}{2} + \tfrac{1}{3} + \tfrac{1}{4} + \ldots + \tfrac{1}{n}$. Then the Riemann hypothesis is equivalent to the claim that

$\sigma(n) \le H_n + e^{H(n)} \log(H_n)$ for all $n \ge 1$, which in turn is
                                            equivalent to the claim that:

$\sigma(n) < e^{\gamma} n \log \log n$ for all $n \ge 5041$

(Lagarias 2002)

squarefree and blue and red numbers

The Riemann hypothesis can also be expressed in terms of the squarefree numbers. Call a number *red* if it is the product of an even number of distinct primes, and *blue* if it is the product of an odd number.

Of the squarefree numbers up to 30, there are eight red numbers, 1, 6, 10, 14, 15, 21, 22, and 26, and eleven blue numbers. The Riemann hypothesis says, very roughly, that the numbers of blue and red squarefree numbers are about the same. More precisely, the Riemann hypothesis is equivalent to this theorem:

> Choose a value of $e > 0$, which can be as small as you like. Then there is a number N such that for all $n > N$, the number of blue numbers from 1 to n inclusive does not differ from the number of red numbers in the same range by more than $n^{1/2+e}$. (The smaller the value of e, the larger N will be.) (Wilf 1987)

Denjoy pointed out that if a squarefree number can be regarded as being equally likely to be blue or red, so that "choosing" its color is like flipping a coin and getting heads or tails, then according to standard statistical theory, if you "toss the coin" n times then the difference between the number of heads and tails as n tends to infinity will grow less rapidly than $n^{1/2+e}$, with probability 1. (Edwards 1974, 268)

Unfortunately, this is only a **heuristic** argument: we could turn it around and say that—okay!—it follows that if the Riemann hypothesis is true, then whether a squarefree number is blue or red is indeed like flipping a coin! Nevertheless, since the latter seems intuitively likely, we can say that our intuitive *feeling* that the Riemann hypothesis is correct is strengthened. Mathematics is by no means only about *logic*!

the Mertens conjecture

In 1897 Franz Mertens (1840–1927) conjectured that for every $n \geq 1$, the difference between the number of red and blue numbers (in the same setup) in the range 1 to n never exceeds $\sqrt{n}$.

In other words, he supposed that we could take $e = 0$ and $N = 1$. This is a much stronger conjecture that was proved false by Odlyzko and te Riele in 1985.

Riemann hypothesis curiosities

Nicolas proved a theorem in 1983 about the values of $\phi(n)$ by first assuming that the Riemann hypothesis is true, and then assuming that it is false—so whatever the final conclusion, his result will be sound!

Although there is much that mathematicians don't know about the zeros of the zeta function, they do know enough to calculate the sum of all their reciprocals, which is, $\frac{1}{2}\gamma + 1 - \log 2 - \frac{1}{2}\log(\pi) =$

0.0230957089 . . . , a formula that brings together γ, π, and (in effect) *e*, the base of natural logarithms. (Finch 2003, 42)

See distributed computing; squarefree numbers

Riesel number

A Riesel number (named after the discoverer of the 18th **Mersenne prime**) is an integer k such that $k \cdot 2^n - 1$ is composite for any integer value of n. Riesel found $k = 509203$ in 1956. The Riesel conjecture is that this is the smallest Riesel number.

right-truncatable prime

The prime 73939133 is the largest prime that repeatedly produces primes when digits are deleted from the right. The numbers 73939133, 7393913, 739391, 73939, 7393, 739, 73, and 7 are all prime.

RSA algorithm

> Can the reader say what two numbers multiplied together will produce the number, 8,616,460,799? I think it unlikely that anyone but myself will ever know for they are two large prime numbers.
>
> —*Stanley Jevons,* Principles of Science *(Gardner 1975, 85)*

The RSA algorithm is the best known method of **public key encryption**, in which the encryption and decryption keys are different. There has to be a connection between the two keys, of course, otherwise they could not work together, but provided this connection cannot be discovered by the fastest available computers in a "reasonable" time, the system is effectively secure. "RSA" is an acronym for Ronald Rivest, Adi Shamir, and Leonard Adleman, who created the algorithm in August 1977. More recently, documents released by GCHQ, the British Government Communications Headquarters, in Britain show that the method was first invented at GCHQ in 1973 by Clifford Cocks, but they did nothing with it. Adleman, Rivest, and

Shamir were less laid-back. They patented the method and formed RSA Data Security Inc. to exploit it commercially.

The simplest "very difficult" problem in number theory is factorizing a large number, so it is no surprise that the RSA method depends on having two large primes, p and q, whose product, $n = pq$, can

---

Adleman originated the term "computer virus" to describe a self-replicating program that could infect a computer. His student Fred Cohen released the first computer virus in 1983, within a small network.

Also in 1983, Adleman created with Carl Pomerance and R. S. Rumely the APR **primality test**.

More recently, at Christmas 1993, Adleman created the first DNA computer. It consisted of twenty-one test tubes of DNA, and he used it to solve a particular case of the Traveling Salesman problem, for which there is no polynomial-time algorithm. (The Traveling Salesman problem asks for the shortest route to visit N towns.)

This is part of Thomas Bass's account of those days:

Rivest and Shamir kept popping off ideas for how to implement a public-key crypto system, and Adleman found the holes. Over the next few months, he cracked 42 potential systems.... One night, Adleman was awakened by a phone call. It was Rivest with public-key crypto system Number 43.... "I knew he'd come up with an unbeatable system," says Adleman. "You get a feel for these things if you think about them long enough; my aesthetic judgment told me he'd finally done it."

Rivest stayed up all night drafting a research paper, which he presented to his colleagues the following morning. Published in the February 1978 Communications of the ACM under the joint authorship of Rivest, Shamir, and Adleman, the paper was officially titled "A Method for Obtaining Digital Signatures and Public-Key Cryptosystems." ... Then, the National Security Agency got in touch. The spies were calling to say that the US government classifies cryptography as a munition, and that if they mailed their article overseas, they would be prosecuted for illegal arms dealing....

Rivest, Shamir, and Adleman later decided to go into the crypto business.... So, how did Adleman shape up as a businessman? He throws back his head and laughs. "In my hands the business went into the toilet," he says. "I was terrible. Just awful. We eventually reorganized the company and hired a real president." (Bass 1995)

safely be made public because n is impossible to factorize with current methods. In 1977 a prime of 100 digits would do. As of today, p and q can safely be of about 200 digits each.

It also depends on **Fermat's Little Theorem**, as improved by Euler. This says that if a and n are coprime, then,

$$a^{w(n)} \equiv 1 \;(\text{mod } n)$$

where $\phi(n)$ is **Euler's totient function**, which in this case, because p and q are distinct primes, equals $(p-1)(q-1)$. Having chosen p and q, we next calculate, using standard techniques, two integers, d and e, such that $1 < d, e < \phi(n)$ and

$$de \equiv 1 \;(\text{mod } \phi(n))$$

The message we wish to send is then enciphered as a number, M, which must be less than n: in practice, the asymmetric RSA system is very slow at encrypting long messages, so they are encrypted by a standard symmetric method and only the symmetric key is encrypted by using the RSA algorithm. The message sent is now, $M^e \;(\text{mod } n)$, using the number e that is made public. The numbers n and e are the *public key*.

The receiver, who has the private value d, deciphers it by calculating $(M^e)^d \;(\text{mod } n)$, since $M^{ed} \equiv M \;(\text{mod } n)$, by Euler's version of Fermat's Little Theorem, and M is less than n.

A cryptanalyst in order to crack the cipher must be able to find M^e given only the values of n and e. This is known as the RSA problem. Although no one has ever proved it, it is widely believed to be equivalent to the factoring problem. If the cryptanalyst can factor n, the cipher can be cracked, otherwise not, and this cannot currently be done in polynomial time.

Martin Gardner's challenge

The RSA system received a burst of publicity in 1977 when the authors published a public key message and offered a modest $100 reward (plus vast kudos) for any crypto-nut who could crack it and Martin Gardner advertised the offer in his *Scientific American* "Mathematical Games" column (Gardner 1977) under the provocative title, "A New Kind of Cipher That Would Take Millions of Years to Break." It didn't, of course, but the 129-digit number known as RSA-129, with the public encryption key = 9007, was not solved until April 26, 1994, by a team led by Arjen Lenstra, who used the Internet, 1,600 assorted computers, and a novel factorization algorithm called the quadratic sieve, for eight months. The message read, "The magic words are squeamish ossifrage."

RSA-129 = 114,381,625,757,888,867,669,235,779,976,146,612,010,218,
296,721,242,362,562,561,842,935,706,935,245,733,897,830,
597,123,563,958,705,058,989,075,147,599,290,026,879,543,
541

RSA Factoring Challenge, the New

If you fancy trying your hand at giant factorizations, the RSA Laboratories (at www.rsasecurity.com) sponsor the RSA Factoring Challenge to "encourage research into large-number factorization and computational number theory."

Eight challenge numbers are posted on the Internet. Each number is a **semiprime**, as used in RSA encryption.

There are prizes for successful factorization, from $10,000 for the smallest 576-bit number to $200,000 for 2048 bits. (If you think you can meet the challenge, you can submit your factorization to RSA Laboratories online for confirmation.)

These are the first two and the last of the current eight challenges. The numbers are designated "RSA-XXXX," where XXXX is the number's length, in bits. The first has already been factored:

challenge	prize	status	date	solver
RSA-576	$10K	factored	12/3/2003	J. Franke et al.
RSA-640	$20K	open		
. . .				
RSA-2048	$200K	open		

RSA-640, with 193 decimal digits, is:

31074182404900437213507500358885679300373460228427
27545720161948823206440518081504556346829671723286
78243791627283380334154710731085019195485290073377 2
48227835257423864540146917366024776523466 09

The prizes offered may sound like a small fortune, but the work involved is immense: these are RSA Laboratories' own estimates of the resources required to factor numbers of various sizes *within one year* (in the second column, "machines" means the number of 500 MHz Pentiums, or similar):

number length (bits)	machines	memory
430	1	trivial
760	215,000	4 Gb
1020	342,000,000	170 Gb
1620	1.6×10^{15}	120 Tb

(RSA Laboratories Bulletin #13)

These data mean that a 512-bit key with 155 decimal digits (which is still used commercially) can be factored in a few hours. Needless to say, improvements in factoring algorithms or in machine hardware could possibly bring larger keys within feasible reach, such as the latest American plans for a superfast computer performing 250 trillion calculations per second.

RSA Laboratories also publishes a Secret-Key Challenge aimed at the U.S. government's data encryption standard (DES), which uses only 56-bit keys. The idea is to use brute force to check all possible keys, from the simplest challenge, a 40-bit key, to the hardest, of 128 bits.

In January 1999, Distributed.net won an RSA Labs contest to crack a message coded using the DES. They took just twenty-four hours using the free time of 100,000 computers worldwide, which tested over 250 billion keys every second. (RSA Laboratories, www. rsasecurity.com)

Ruth-Aaron numbers

Hank Aaron hit his 715th Major League home run on April 8, 1974, beating the record of 714 held by Babe Ruth. Carl Pomerance, inventor of the quadratic sieve factorization method, and two colleagues noticed that the pair 714 and 715 have some curious and interesting properties.

The number $714 = 2 \times 3 \times 7 \times 17$, and $715 = 5 \times 11 \times 13$; so $714 \times 715 = 2 \times 3 \times 5 \times 7 \times 11 \times 13 \times 17$, the product of the first seven primes, or 7 **primorial**.

They then discovered on a computer that only primorials 1, 2, 3, 5, and 7 can be represented as the product of consecutive numbers, up to primorial 3049.

They also noticed that $\sigma(714) = 1728 = 12^3$, while the ratio $\sigma(714)/\phi(714) = 1728/192 = 9$, a perfect square.

We will add these additional properties:

- $2 + 3 + 7 + 17 = 5 + 11 + 13 = 29$, and this is the only way to split the first seven primes into two sets of equal total.
- $\phi(\phi(715) = \phi(480) = 128$, which is double $\phi(\phi(714) = \phi(192) = 64$.

(Nelson, Penney, and Pomerance 1974)

Scherk's conjecture

H. F. Scherk conjectured on the basis of empirical evidence that the nth prime, if n is even, can be represented by the addition and subtraction of all the smaller primes, each taken once. For example, 13 is the 6th prime, and

$$13 = 1 + 2 - 3 - 5 + 7 + 11$$

Similarly, the nth prime, when n is odd, can be represented under the same conditions, except that the immediately smaller prime is doubled:

$$17 = 1 + 2 - 3 - 5 + 7 - 11 + 2 \cdot 13$$

The conjecture was proved in 1967 by J. L. Brown.

Chris Nash has proved that every integer N (greater than 17) is an algebraic sum of all the primes less than N, or all the primes less than N except for the last. (Nash 2000)

Both Brown and Nash used **Bertrand's postulate**.

semiprimes

A semiprime (or *semi-prime*) or **2-almost-prime** has just two prime factors. The sequence starts, 4, 6, 9, 10, 14, 15, 21, 22, 25 26 33 34 35 38 39 46 49 51 55, . . . (Sloane A001358)

- There are 299 semiprimes less than 1000.
- All semiprimes are **deficient**, except for 6, which is **perfect**.

Note that all the semiprimes from 9 to 39 form pairs or triplets of consecutive semiprimes. The next such pairs are 57-58, 85-86-87, 93-94-95, 118-119, 122-123, 133-134, 141-142-143, 145-146.

If the prime factors are required to be distinct, then the first consecutive pair is 14 and 15; the first consecutive pair with three distinct prime factors each is $230 = 2 \cdot 5 \cdot 23$ and $231 = 3 \cdot 7 \cdot 11$.

Four consecutive semiprimes cannot exist, because one of them would be divisible by 4 and have at least three factors.

J. R. Chen proved that every even number is the difference between a prime and a semiprime.

sexy primes

Sexy primes are such that n and $n + 6$ are both prime. ("Sexy" comes from the Latin for "six," which as all schoolboys know, is *sex*.) The sexy pairs under 200 are: 5-11, 11-17, 13-19, 17-23, 23-29, 31-37, 37-43, 47-53, 53-59, 61-67, 67-73, 73-79, 83-89, 97-103, 103-109, 107-113, 131-137, 151-157, 157-163, 167-173, 173-179, and 191-197. (Sloane A023201) There are ten of the form $6n + 1$ and twelve of the form $6n - 1$.

The sequence of sexy triplets starts: 7-13-19, 17-23-29, 31-37-43, 47-53-59, . . . (Sloane A046118)

The sexy quadruplets are: (5-11-17-23), 11-17-23-29, 41-47-53-59, 61-67-73-79, . . . (Sloane A023271) The first quadruplet is in parentheses because it is the only one that does not start with a unit digit 1.

The sexy quintuplet, 5-11-17-23-29, is unique since one of the numbers must be divisible by 5. There is, ironically, no sexy sextuplet.
    See twin primes; cousin primes

Shank's conjecture

The gap between succesive primes p_n and p_{n+1} is taken to be $p_{n+1} - p_n - 1$, so that the gap between 7 and 11 is 3, corresponding to the composites 8, 9, and 10. Shank's conjecture is that if $p(g)$ is the first prime that follows a gap of size g, then $\log p(g)$ is roughly $\sqrt{g}$.

Siamese primes

Named by Beauregard and Suryanarayan, they are prime pairs of the form $n^2 - 2$ and $n^2 + 2$. The sequence of pairs starts, 7-11, 79-83, 223-227, 439-443, 1087-1091, 13687-13691, . . . (Beauregard and Surya-narayan 2001)

Sierpinski numbers

Named after Waclav Sierpinski (1882–1969), a Sierpinski number is an integer k such that $k \cdot 2^n + 1$ is composite for *any* integer value of n. Sierpinski proved in 1960 that they exist. In 1962, John Selfridge discovered the smallest known Sierpinski number, 78557. Whatever the value of n, $78557 \cdot 2^n + 1$ is divisible by one of the primes 3, 5, 7, 13, 19, 37, or 73.

The next largest is 271129. However, there are several "probable Sierpinski numbers," including 4847, which produces composite numbers up to $n = 279700$. (Rivera, Problem 30)

Sierpinski strings

Sierpinski proved that if you take any string of decimal digits, such 1449487, then there are infinitely many prime numbers starting with this string. There is also an infinity of prime numbers *ending* with any given string, provided the last digit is 1, 3, 7, or 9.

This is also true of any arithmetic progress, $an + b$, provided a and b are coprime.

Sierpinski's quadratic

Sierpinski proved in 1964 that for every positive integer k, there is an integer b such that $n^2 + b$ is prime for at least k values of n.

Sierpinski's $\phi(n)$ conjecture

Sierpinski conjectured that for every integer $k \geq 2$, there is a number m for which $\phi(n) = m$ has exactly k solutions. This was proved by Kevin Ford in 1999. (Ford 1999)

See Riesel number

Sloane's *On-Line Encyclopedia of Integer Sequences*

There have been *tables* of assorted numbers, including the primes, for centuries, but, as Neil Sloane points out, until the publication of his book *A Handbook of Integer Sequences* in 1973, there was no way to find out whether a particular sequence of integers was well known and, if so, what it represented.

That book was so popular that a second, enlarged edition, *The Encyclopedia of Integer Sequences*, appeared in 1995, in which Sloane and Simon Plouffe described nearly 6,000 examples. Now the *On-Line Encyclopedia* (www.research.att.com/~njas/sequences) has more than 93,000 sequences, many contributed by Sloane's several hundred correspondents. Each entry contains the leading terms of the sequence, keywords, mathematical motivations, literature links, and more.

The great feature of the books and the *On-Line Encyclopedia* is that they are arranged in "numerical" order. Online you can simply enter the first few terms of a sequence to locate it—if it is recognized! (You can also check by keyword, which is also invaluable.)

Readers will recognize sequence A000040, which is the prime numbers. This entry gives a definition and then lists twenty-two references to papers and books and nearly fifty links to other sites, as well as eighteen other related Sloane encyclopedia sequences.

However, if you type in the sequence of integers, 2, 3, 5, 7, 11, 13, 17, 19, 23, and expect to get only the answer "The prime numbers," you will be greatly mistaken. Several other sequences start the same way, only to diverge sooner or later—sometimes much later. Some examples:

- A005180 starts the same, but is "The order of the simple groups."
- A008578 includes an initial 1, and is "The prime numbers at the beginning of the 20th century (today 1 is no longer regarded as a prime)."
- A030059 is the "Product of an odd number of distinct primes."
- A015919 is "Numbers n such that $n \mid 2^{n-2}$," so it includes all the primes plus the pseudoprimes to base 2, starting with the smallest, 341.

- A038179 is more mysterious. It starts 2, 3, 5, 7, 11, 13, 17, 19, 23, 25, 29, 31, 35, 37, 41, 43, 47, 49, 53, 55, 59, 61, . . . and is the "Result of second stage of sieve of Eratosthenes." This entry is followed by the results of the third and fourth stages.
- The next, A049551, is "Primes p such that $x^{19} = 2$ has a solution mod p."

Higgs primes, "prime-like" sieves using **Fibonacci numbers**, primes with digits in ascending order . . . the list is not endless but it does go on and on, and you can examine it at your leisure, thanks to Neil Sloane!

Smith numbers

Albert Wilansky named Smith numbers after his brother-in-law, whose telephone number, $4937775 = 3 \cdot 5 \cdot 5 \cdot 65837$, has the property that its digits' sum is the same as that of its prime factors, 42. Another example is: $666 = 2 \cdot 3 \cdot 3 \cdot 37$.

This is an excellent example of a property, spotted by chance, that is plausibly a bit of **trivia**: so much so that at least one article has been published by an indignant author arguing that Smith numbers are indeed a waste of time and don't deserve the effort devoted to them.

In particular, since they depend entirely on the base being used, aren't they surely superficial? "Those who investigate Smith numbers are not trying to penetrate deep into the secrets of integers. They are instead observing mere accidents of their representation in an arbitrary system," according to Underwood Dudley. (Ivars Peterson's *MathTrek*, October 27 1997: www.maa.org/mathland) (Underwood 1994)

Maybe—but could they not be studying deep properties of certain bases that allow Smith numbers to exist at all? This disagreement illustrates how mathematicians, amateur and professional, differ in their judgments of what is worthwhile—and sure enough, Smith numbers turn out to be not totally devoid of interest.

All primes are Smith numbers, uninterestingly, and so are, 4, 22, 27, 58, 85, 94, 121, . . . and the number of the beast, 666. Much larger isolated examples are known. Samuel Yates found a giant of 13,614,513 digits.

There is also an ingenious method of creating Smith numbers, which links them to the prime **repunits**. Sham Oltikar and Keith Wayland found in 1983 that if R_n is prime then $3304 \cdot R_n$ is a Smith number. Moreover, factor 3304 can be replaced by any of these numbers: 1540, 1720, 2170, 2440, 5590, 6040, 7930, 8344, 8470, 8920, . . .

There are 376 Smith numbers less than 10,000 and 29,928 Smiths less than 10^6. It is conjectured that about 3% of any million consecutive integers are Smith numbers. (*Wikipedia* online encyclopedia, Smith number)

If there is an infinity of prime repunits, as conjectured, then there will be an infinity of non-trivial Smith numbers. Wayne McDaniel has since proved (1987) that there is indeed an infinity of Smith numbers, without resolving that conjecture.

Another method is to start with any prime whose digits are all zeros and 1s and multiply it by a suitable factor. Here are some of Hoffman's examples:

$$101 \cdot 2 = 202$$
$$10111 \cdot 140 = 1415540 = 10111 \cdot 2 \cdot 2 \cdot 5 \cdot 7$$
$$101111 \cdot 21 = 2123331 = 101111 \cdot 3 \cdot 7$$

(Hoffman 1998)

Pat Costello in 1984 published seventy-five Smith numbers of the form $p \cdot q \cdot 10^k$ where p is a small prime and q is a **Mersenne prime**. His largest example was $191 \cdot (2^{216091} - 1) \cdot 10^{266}$, with 65,319 digits.

Kathy Lewis in 1994 found an infinite sequence of Smith numbers of the form $11^a \cdot 9R_n \cdot 10^b$, where R_n is the nth repunit. (Walter Schneider, www.wschnei.de: "Smith Numbers")

Smith brothers

The smallest consecutive Smith brothers are 728-729, 2964-2965, 3864-3865, . . .

See Rhonda numbers; trivia

smooth numbers

Smooth numbers are very composite numbers, having many small prime factors. They are a kind of dual to the prime numbers—a prime has one big prime factor, itself, and smooth numbers have none.

A number is *k*-smooth if it has no prime factor greater than *k*. So the powers of 2 are the only 2-smooth numbers, and the 3-smooth numbers are of the form 2^n3^m, and the 5-smooth numbers are of the form $2^n3^m5^q$, and so on.

There are forty-six 10-smooth numbers less than 100, 140 less than 1000, 332 less than 10,000, and 587 less than 100,000.

Carl Pomerance used smooth numbers to prove that there are an infinite number of **Carmichael numbers**, and they are also used in primality testing and factorization algorithms that often depend on finding smooth numbers within a given range.

See [Ramanujan's] highly composite numbers

Sophie Germain primes

Sophie Germain (1776–1831) was one of the very earliest female mathematicians. She educated herself at home in the library of her father, who became a director of the Bank of France, and as a thirteen-year-old she read of the death of Archimedes, killed by a Roman soldier as he examined a figure traced in the sand, and decided to become a mathematician. At the age of eighteen, she obtained the notes for Lagrange's lectures on analysis, posed as a student using the name Le Blanc, and sent a paper to Lagrange. Extremely impressed, he decided to meet the young author, and so discovered that "he" was a woman. They corresponded, and he included some of her results in his *Theory of Numbers*, giving her credit in a footnote.

A little later, she also wrote to Gauss under the same pseudonym, having studied his *Disquisitiones Arithmeticae*, and he was also astonished when he discovered the sex of his correspondent, remarking that "Sophie Germain proved to the world that even a woman can accomplish something in the most rigorous and abstract of sciences."

Her greatest achievement during her lifetime was to win the prize of the Académie des Sciences in response to a challenge to explain the creation of Chladni figures (seen when sand is scattered on a vibrating plate), but she is best remembered today for the Sophie Germain primes. These are the odd primes q for which $2q + 1$ is prime, too. She proved in 1823 if p is a Sophie Germain prime, then there are no integers, x, y, and z, none zero, and not multiples of p, such that $x^p + y^p = z^p$. This is an example of the "first case" of **Fermat's Last Theorem**.

The sequence of primes, p, such that $2p + 1$ is also prime, starts,

> 2, 3, 5, 11, 23, 29, 41, 53, 83, 89, 113, 131, 173, 179, 191, 233, 239, 251, 281, 293, 359, . . . (Sloane A005384)

These are also the primes for which $\phi(n)$ is double a prime. Triplets such that p, $2p + 1$, and $4p + 3$ are all prime start with these values of p:

> 2, 5, 11, 41, 89, 179, 359, 509, 719, 1019, 1031, 1229, 1409, . . . (Sloane A007700)

The three largest known Sophie Germain primes are:

Sophie Germain prime	no. of digits	year discovered
$2540041185 \cdot 2^{114729} - 1$	34,547	2003
$18912879 \cdot 2^{98395} - 1$	29,628	2002
$1213822389 \cdot 2^{81131} - 1$	24,432	2002

(Caldwell, *Prime Pages*)

safe primes

Sophie Germain primes are related to cryptography. If p and $2p + 1$ are prime, then $q = 2p + 1$ is said to be a safe prime, because $q - 1$ does *not* have many small factors and therefore cannot easily be factored, making the encryption more secure.

squarefree numbers

Many integers can be written as the product of a perfect square and a number with no squared factor. For example,

$$12 = 2^2 \cdot 3 \qquad 50 = 5^2 \cdot 2$$
$$288 = 12^2 \cdot 2 \qquad 35 = 5 \cdot 7$$

Other integers are *squarefree*. So 35 is squarefree, as are these numbers that are each the product of distinct prime factors, apart from 1:

$$1, 2, 3, 5, 6, 7, 10, 11, 13, 14, 15,$$
$$17, 19, 21, 22, 23, 26, 29, 30, \ldots$$

The proportion of squarefree numbers less than N is roughly constant, and tends to cN, where $c = 6/\pi^2$. Another way of putting this is: what is the probability that a number is squarefree, having no repeated prime factor?

The number of squarefree numbers less than n is equal to $6n/\pi^2$ plus a factor that is roughly proportional to $\sqrt{n}$ as n tends to infinity. On the other hand, the average number of square divisors of a number N is roughly $\pi^2/6$. (Greger 1978)

The probability that two integers are coprime, that is, prime to each other, is also $6/\pi^2$. Both probabilities turn out to be equal to $1/\zeta(2)$, where $\zeta(n)$ is the zeta function introduced by Euler.

Stern prime

A Stern prime is a prime *not* of the form $p + 2a^2$ where p is a prime and $a > 0$. The largest known Stern prime is 1493. (Russo: Caldwell, *Prime Pages*)

strong law of small numbers

Many mathematical sequences start by showing what seems to be a strong pattern—but then the pattern disappears! The **Fermat numbers** are a perfect example. The first five are prime, but then the pattern collapses.

The prime **race** between $4n + 1$ and $4n + 3$, analyzed by **Littlewood**, is a more subtle example—it is a long time before the lead changes at all, though eventually it changes infinitely often.

If we calculate powers of $3/2$ and then take the integral part, we get this sequence, which is also suggestive:

n	1	2	3	4	5	6
$(3/2)^n$	1.5	2.25	3.375	5.0625	7.59375	11.390625
	1	2	3	5	7	11

Well, well! Can this be the sequence of primes? No, of course not, because (apart from the 1 at the start) the next number will be much greater than 11. In fact it is 17, which is prime, but then the sequence continues, 25, 38, 57, 86, . . .

That example comes from Professor Richard Guy, who for many years edited the "Problems" section of the *American Mathematical Monthly* and is also the author of *Unsolved Problems in Number Theory* (1994). He has discussed this tricky phenomenon under the title "The Strong Law of Small Numbers." (Guy 1988; and 1990) As he puts it,

> "There aren't enough small numbers to meet the many demands made of them."

> *or* "Capricious coincidences cause careless conjectures."

> *and* "Early exceptions eclipse eventual essentials."

Here are some of his amusing examples involving prime numbers:

The numbers 31, 331, 3331, 33331, 333331, and 3333331 are all prime, and so is the next number in the sequence, but 333333331 = 17 · 19607843. Early numbers in the sequence are likely to be prime because no number in the sequence is divisible by 2, 3, 5, 7, 11, 13, or 37.

The alternating sums of the **factorials** seem to be always prime:

$$3! - 2! + 1! = 5$$
$$4! - 3! + 2! - 1! = 19$$
$$5! - 4! + 3! - 2! + 1! = 101$$
$$6! - 5! + 4! - 3! + 2! - 1! = 619$$
$$7! - 6! + 5! - 4! + 3! - 2! + 1! = 4421$$

The next in the sequence is also prime, but $9! - 8! + 7! - 6! + 5! - 4! + 3! - 2! + 1! = 79 \cdot 4139$.

Can we always be certain, however, that a phenomenon is an example of the strong law of small numbers? No, the mathematician can only make a judgment.

If you divide 2^n by n (from $n = 1$ onwards), this is how the sequence of *remainders* starts:

n	1	2	3	4	5	6	7	8	9	10
2^n	2	4	8	16	32	64	128	256	512	1024
remainder	0	0	2	0	2	4	2	0	8	4

The remainders appear to always be powers of 2. Are they? The remainder sequence continues, 2, 4, 2, 4, 8, 0, 2, but then we come to $2^{18} = 262144$ and $262144 \equiv 10 \pmod{18}$.

The sequence then continues, however, 2, 16, 8, 4, 2, 16, but then breaks down again: $2^{25} = 33554432 \equiv 7 \pmod{25}$.

Unlike the previous examples, we can't say that the pattern has broken down completely. Rather, we might look for patterns among the exceptions. For example, how many odd remainders are there? Are all odd numbers a remainder *eventually*? When does the first remainder 3 occur, if ever? As it happens, D. H. and Emma Lehmer discovered that $2^n \equiv 3 \pmod{n}$ for the first time when $n = 4700063497$.

By sheer chance, it is not just likely but certain that there will be coincidental patterns among the vast numbers of functions that a modern computer can calculate. So we should be suspicious when it turns out that $21 \cdot 2^n - 1$ and $7 \cdot 4^n + 1$ are both prime when $n = 1, 2, 3, 7, 10$, and 13, and composite for other values up to 17. The match then collapses. The first function is prime for $n = 18$ while the second is composite, and we can be pretty confident that the "pattern" was a statistical hoax.

Here is one more example, not taken from Richard Guy's articles: add an even number of consecutive primes, from 2, and then add 2:

$$
\begin{aligned}
2 + 3 + 2 &= 7 \\
2 + 3 + 5 + 7 + 2 &= 19 \\
2 + 3 + 5 + 7 + 11 + 13 + 2 &= 43 \\
2 + 3 + 5 + 7 + 11 + 13 + 17 + 19 + 2 &= 79 \\
2 + 3 + 5 + 7 + \ldots + p_{2n} + 2 &=
\end{aligned}
$$

The results *seem* always to be prime, but does this pattern continue? No—the sequence breaks down when $p_{2n} = p_{22}$ and,

$$2 + 3 + \ldots + 79 + 2 = 793 = 13 \cdot 61$$

On the other hand, the sums taking an odd number of primes as far as p_{2n+1} are not only composite, being even, but tend to have several factors. The first few are 4, 12, 30, 60, 102, 162, and 240.

Mathematics, and especially number theory, and not least the prime numbers, are indeed exceptionally full of patterns, but our very human capacity to spot patterns is even more exceptional!

See conjectures; Euler's totient (phi) function; induction

triangular numbers

The sequence $\frac{1}{2}n(n + 1)$ and the number of dots in a triangular array:

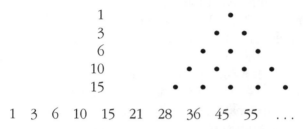

1
3
6
10
15

1   3   6   10   15   21   28   36   45   55   ...

(Sloane A000217)

Thomas Greenwood noticed that 1 more than an even or 2 less than an odd triangular number is often a prime number.

— 1   7   11   13   19   29   37   43   53   ...

$T_{31} = 496$ is the first counterexample; 31 is prime but 7 divides 497.

Every even **perfect number** is triangular.
See Pascal's triangle

trivia

G. H. **Hardy** in his book *A Mathematician's Apology* gave several examples of "trivial" properties of numbers, including the fact that 153 is the sum of the cubes of its digits,

$$153 = 1^3 + 5^3 + 3^3$$

and remarked, "These are odd facts, very suitable for puzzle columns and likely to amuse amateurs, but there is nothing in them which appeals to the mathematician."

As if to endorse Hardy's view, I notice that the current edition of *A Mathematician's Apology* has exactly 153 pages. So what? There are indeed many recorded properties of numbers that seem little, if at all, more significant than traditional number mysticism. An iccanobiF prime is a **Fibonacci number** reversed, for example, 31 or 773. Is that trivial or what? We can sink even lower. A James Bond prime has been defined as any prime ending in—you guessed it!—007! The first three are 4007, 6007, and 9007, but **Sloane's *On-Line Encyclopedia of Integer Sequences*** declines to include it, naturally.

The five-digit prime 12421 has a different property: can you spot it? Yes, the digits rise and then fall, making it a *Mountain prime*, as well as palindromic. I'm surprised it isn't called a Peaky prime. It could then be referred to euphoniously as a Perfectly Palindromic Peaky Prime. But so what?

The largest known Holey prime is 4 followed by 16,131 9s. It's Holey because the digits 4 and 9 (and 0, 6, and 8, which do not feature) have *holes* inside, as written in our current Arabic numerals. Actually, the devotees of Holey-ness have competition, because a Pime has been defined as a prime that uses only digits with closed loops, which means 0, 6, 8, and 9 but *not* 4. The smallest Pime is 89—now you know—but the second Holey prime, 409, is not a Pime though the second Pime is of course Holey, because *all* Pimes are Holey.

A Yarborough prime is defined rather differently but with similar effect: its only digits are 2, 3, 4, 5, 6, 7, 8, or 9, so an anti–Yarborough prime is—yes!—a prime made up of zeros and 1s (and not just in binary). I can't help feeling that "Holey Yarborough Pimes!" sounds like an obscure swear word.

I do apologize. I am being carried away by the Spirit of Triviality. To be solemn for a moment, it is indeed a delicate issue to decide when trivialness turns into seriosity. For example, **Smith numbers** have often been dismissed as trivial because their definition depends on their digits—but there is a well-known connection between Smith numbers and **repunits**, and no one thinks that repunits are trivial, because they are closely related to **decimals** and their periods (there go the digits again!) and there are some deep unsolved problems and conjectures about decimal periods, including **Artin's conjecture**, named after one of the greatest mathematicians of the twentieth century. QED! Smith numbers are non-trivial.

Now, what about $153 = 1^3 + 5^3 + 3^3$? Any offers?

See also Hardy; Smith numbers

twin primes

The primes include many pairs differing by 2: 3-5, 5-7, 11-13, and 17-19, and so on. These are the twin primes. The only number that appears in two twin pairs is 5, and 3 is the only number in the sequence not of the form $6n \pm 1$.

The number of twin primes less than N naturally declines as N increases. The numbers of pairs less than 10^n are,

2, 8, 35, 205, 1224, 8169, 58980, 440312, 3424506, 27412679, 224376048, . . . (Sloane A007508)

The twin primes conjecture is that there is an infinity of them. Hardy and Littlewood went further and conjectured that the number of twin primes less than n is approximately $2Cn/(\log n)^2$, where $C = 0.6601618158$ . . . is the *twin prime constant*.

Provided x is greater than a certain integer, x' (which can in theory be calculated) then

$$\pi_2(x) < \frac{100x}{(\log x)^2}$$

where $\pi_2(x)$ is the number of pairs of twin primes less than x. (Brun 1920) It was Brun who proved that the sum of the reciprocals of the twin primes converges.

Jing-Run Chen proved in 1966 that there is an infinity of primes p such that $p + 2$ is either prime or the product of two primes. If p is prime and $p + 2$ is either a prime or a **semiprime**, then the number of such pairs less than n is at least $1.05 \times 2Cn/(\log n)^2$ where C is the same as before.

The current record twin primes are $33218925 \cdot 2^{169690} \pm 1$, with 51,090 digits. (Papp, Jobling, Woltman, and Gallot 2002) This is well ahead of the preceding four records:

twin primes	no. of digits
$60194061 \cdot 2^{114689} \pm 1$	34,533
$1765199373 \cdot 2^{107520} \pm 1$	32,376
$318032361 \cdot 2^{107001} \pm 1$	32,220
$1807318575 \cdot 2^{98305} \pm 1$	29,603

(Caldwell, *Prime Pages*)

twin curiosities

- The pair 659-661 is the start of a record-breaking gap between twin primes: the next pair is 809-811.
- All primes except 2 and 3 are of form $6n \pm 1$, so all twin

primes apart from 3-5 are of the form $6n \pm 1$ also. Note that $120 = 6 \cdot 20$ is the smallest integer such that neither $6n + 1$ nor $6n - 1$ is prime.

- The pair $60n^2 + 30n - 30 \pm 1$ is a twin pair for $n = 1$ to 13. (Blanchette: Caldwell, *Prime Pages*)
- Among primes of the form $n^4 + 1$ there are a surprising number of "twins" for which $n^4 + 1$ and $(n + 2)^4 + 1$ are both prime.
- This is the sequence of n for which $n^4 + 1$ is prime: the twins are in bold:

1, **2, 4, 6**, 16, 20, 24, 34, **46, 48, 54, 56**, 74, **80, 82, 88, 90**, 106, 118, 132, **140, 142**, 154, 160, 164, 174, 180, 194, 198, . . . (Lal 1967)

- Thomas R. Nicely has calculated (2004) that the number of pairs of twin primes less than $5 \cdot 10^{15}$ is 5,357,875,276,068. (Nicely 2004a)
- Nicely has also calculated the number of prime quadruplets of the form p, $p + 2$, $p + 6$, and $p + 8$ to the same limit: it is 13,725,978,764. (Nicely 2004b)
- The smallest triplet is 5-7, 11-13, 17-19.
- Carlos Rivera has also found these three examples of triples: 4217-4219, 4229-4231, 4241-4243 with common difference 12; the three pairs starting 208931-208933 with common difference 30; and the three pairs starting 263872067-263872069 with common difference 30. Phil Carmody (2001) found the triplet starting 127397154761-127397154763, also with common difference 30. (Rivera, Puzzle 122) J. K. Andersen has found these two quadratics such that $f(x) \pm 1$ is a pair of twin primes for $x = 0$ to 15:

$$f(x) = 4515x^2 - 67725x + 603900$$
$$f(x) = 12483x^2 - 187245x + 834960$$

(Rivera, Problem 44)

Stop Press!

MathWorld Headline News at mathworld.wolfram.com/news has just reported (July 9, 2004) that R. F. Arenstorf may be close to settling the twin primes conjecture. A hole has been found in his "proof" (remember Andrew Wiles and the gap in his proof of **Fermat's Last Theorem** . . .) but mathematicians are hopeful that it can be filled.

See Brun's constant; Hardy-Littlewood conjectures; Mertens constant

Ulam spiral

Stanislav Ulam (1909–1984) discovered, or invented, his spiral in 1963, while sitting through a boring talk. His doodle looked like this:

He noticed at once that the primes seemed to settle onto certain diagonal lines, and realized that this was because squares lying on straight lines have differences that are increasing linearly—so they represent quadratics. Hence, "It is a property of the visual brain which allows one to discover such lines at once and also notice many other peculiarities of distribution of points in two dimensions," as Ulam and his coworkers remarked in a 1964 paper published in the *American Mathematical Monthly*. (Stein, Ulam, and Wells 1964) In this figure the main diagonal is the function $n^2 + n + 1$. The next figure starts with 41 and the values of **Euler's quadratic** formula appear along the marked diagonal.

The deeper problem is why some quadratic expressions produce such a high proportion of primes. Among the values of Euler's prime quadratic $x^2 + x + 41$, of the first 2,398 numbers generated by the for-

	77	76	75	74	73	72	71
	78	57	56	55	54	53	70
	79	58	45	44	43	52	69
	80	59	46	41	42	51	68
	81	60	47	48	49	50	67
	82	61	62	63	64	65	66
	83	→					

mula, exactly half are primes. Checking all such numbers less than 10,000,000, Ulam and his coworkers found the proportion of primes to be 0.475.

For the quadratic $4x^2 + 170x + 1847$, the proportion of primes is 0.466; for $4x^2 + 4x + 59$, it is 0.437. Other quadratics go to the other extreme: only 5% of the products of the formula $2x^2 + 4x + 117$ are prime.

See Euler's quadratic; and the reference in the "Some Prime Web Sites" section of the bibliography to Dario Alpern's Ulam spiral applet

unitary divisors

If $N = st$ and if s and t have no common factor, then s and t are unitary divisors of N. By convention, 1 and N are included. So the unitary divisors of $24 = 3 \cdot 2^3$ are 1, 3, 8, and 24 only.

All the divisors of a number are unitary, if and only if it is a prime or the product of distinct primes.

Euler's generalization of Fermat's Little Theorem says that if $p \nmid a$, then $a^{\phi(p)} \equiv 1 \pmod{p}$. It is also true that if a is a unitary divisor of n, then there is a number, k, greater than 1, such that $a^k \equiv a \pmod{n}$. For example, 8 is a unitary divisor of 24, although 8 and 24 are not coprime, and $8^3 \equiv 8 \pmod{24}$.

New "World Record" Unitary Aliquot Sequence Completed

by Jack Brennen

On 11 July 2001, the longest known terminating unitary aliquot sequence was computed to its termination.

A unitary aliquot sequence is constructed similarly to a "standard" aliquot sequence, except that instead of adding divisors, one adds up unitary divisors. A unitary divisor D of a number N satisfies two requirements: D divides evenly into N, and D and N have no common divisor other than 1.

It is conjectured that all unitary aliquot sequences eventually either enter a closed cycle, or terminate by reaching the number 1. Certainly no counterexample is known, and the author has now verified the conjecture for all starting numbers up to 400000000.

This new "world record" unitary aliquot sequence begins with the number 151244562, and terminates 16657 steps later by reaching the number 1.

The largest member of the sequence occurs at step 4641, and is a 90-digit number: . . .

$$C90 = 2 \cdot 3 \cdot 3 \cdot 3 \cdot 3 \cdot 503 \cdot 9682217399 \cdot P76$$

$$P76 = 1252135446163593391182747087802437681221287268786627017667109514519500763883$$

In all, there were 40488 distinct prime factors needed to complete this sequence, and thus 40488 numbers were proven prime. . . .

The computation of the sequence occurred over a period of about 29 weeks, using one 450 MHz Pentium-II on a full-time basis, and three other similar machines on a part-time basis. . . .

(Jack Brennen, www.brennen.net/primes/)

---

unitary perfect

If n is the sum of all of its unitary divisors it is unitary perfect. Unitary perfects are naturally rarer than ordinary perfect numbers. The first five are: 6, 60, 90, $87360 = 2^6 \cdot 3 \cdot 5 \cdot 7 \cdot 13$ and

$$2^{18} \cdot 3 \cdot 5^4 \cdot 7 \cdot 11 \cdot 13 \cdot 19 \cdot 37 \cdot 79 \cdot 109 \cdot 157 \cdot 313$$

See aliquot sequences (sociable chains); amicable numbers

untouchable numbers

An untouchable number is any number that cannot be the sum of aliquot parts (proper **divisors**) of n. In other words, it is never a value of $\sigma(n) - n$. The sequence starts, 2, 5, 52, 88, 96, 120, 124, 146, 162, 178, . . . (Sloane A005114)

The only known odd untouchable is 5, and it is conjectured to be unique.

There is an unfinity of untouchable numbers. (Erdös 1973)

The numbers of untouchables less than 10^n for $n = 1, 2, 3, 4 \ldots$ start, 2, 5, 89, 1212, . . . (Sloane A057978)

weird numbers

A number is weird if it is **abundant** but not **pseudoperfect**. The smallest weird number is 70, because its factors sum to 74:

$$1 + 2 + 5 + 7 + 10 + 14 + 35 = 74$$

but no subset of these factors sums to 70.

There is an infinite number of weird numbers, including twenty-four under 10^6. The sequence starts,

70, 836, 4030, 5830, 7192, 7912, 9272, 10430, 10570, 10792, 10990, 11410, 11690, . . . (Sloane A006037)

Wieferich primes

These are named after Arthur Joseph Alwin Wieferich (1884–1954), who in 1909 published a surprising criterion for the *first case* of **Fermat's Last Theorem**: if there exist integers x, y, z such that $x^p + y^p + z^p = 0$ where p is an odd prime and p does not divide xyz, then $p^2 \mid 2^{p-1} - 1$. Such primes are called Wieferich primes. (**Fermat's Little Theorem** says that if p is prime, then it divides $2^{p-1} - 1$.)

236 • Wilson's theorem

The only known Wieferich primes are 1093, discovered by W. Meissner in 1913, and 3511, discovered by N. G. W. H. Beeger in 1922. (Sloane A001220)

Richard McIntosh completed a Wieferich prime search, on March 9, 2004, up to $1.25 \cdot 10^{15}$, without finding any new primes.

The two known Wieferich primes lead to these curious digit patterns in binary: $1092_2 = 10001000100$; $3510_2 = 110110110110$.

In 1910, Mirimanoff added that if the first case of Fermat's Last Theorem is false for exponent p, then $3^{p-1} - 1$ is divisible by p^2.

The **abc conjecture** implies that there exists an infinity of Wieferich primes. The number less than x is at least $C \log x$ for some constant C. However, Silverman also proved, in 1988, that if the abc conjecture is true, then for any a greater than 1 there is an infinity of primes p, for which $p^2 \nmid a^{p-1} - 1$.

The two values of p between 3 and 2^{32} for which $p^2 \mid 3^{p-1} - 1$ are 11 and 1006003.

Curiously, when $a = 99$, there are five small values of p for which $p^2 \mid a^{p-1} - 1$: $p = 5, 7, 13, 19$, and 83. (Montgomery 1993)

It has been conjectured that the probability that a prime p is a Wieferich prime is $1/p$. It has also been conjectured that **Mersenne numbers** with prime exponents are **squarefree**. This is made highly probable by the theorem that if $p^2 \mid M_q$, then p is a Wieferich prime.

The condition p^2 divides $a^{p-1} - 1$ for $a = 5$, has solution $p = 188748146801$, and the "reverse" is true also: $p^{a-1} - 1$ is divisible by $a^2 = 25$. (Keller and Richstein 2004)

Wilson's theorem

John Wilson (1741–1793) is only remembered for Wilson's theorem, which states that if p is prime, then $(p - 1)! + 1$ is divisible by p, as illustrated by the first few primes:

p	2	3	5	7	11
$(p-1)! + 1$	2	3	25	721	3,628,801 (= 11 · 329,891)

Ironically, Wilson was not only not the first to discover this theorem—it appears in Leibniz's papers—but he didn't prove it. Lagrange did that in 1773, and also showed the converse, that if the equation is true, then p is prime.

Wilson's theorem can be expressed in different forms. For example:

$$(p-2)! - 1 \equiv 0 \pmod{p}$$
$$2(p-3)! + 1 \equiv 0 \pmod{p}$$
$$6(p-4)! - 1 \equiv 0 \pmod{p}$$

and in general: $(q-1)!(p-q)! - (-1)^q \equiv 0 \pmod{p}$

It follows from Wilson's theorem that if p is an odd prime, then

$$1^2 \cdot 3^2 \cdot 5^2 \dots (p-2)^2 \equiv (-1)^{(p+1)/2} \pmod{p}$$

and $\qquad 2^2 \cdot 4^2 \cdot 6^2 \dots (p-1)^2 \equiv (-1)^{(p+1)/2} \pmod{p}$

Wilson's theorem is a very inefficient way of proving that a number is prime: the largest ever proved by its use is probably 1099511628401. (Rupinski: Caldwell, *Prime Pages*)

twin primes

Wilson's theorem implies that n and $n + 2$ are a pair of twin primes if and only if,

$$4((n-1)! + 1) + n \equiv 0 \bmod n(n+2)$$

For example, if $n = 5$, $4((n-1)! + 1) + n = 105 = 3 \cdot 5 \cdot 7$.

Wilson primes

A Wilson prime is a prime number such that $(p-1)! \equiv -1 \pmod{p^2}$. The only known Wilson primes are 5, 13, and 563, found by Goldberg in 1953, using an early electronic computer. There are no others less than 500,000,000.

By a **heuristic** argument, the 4th Wilson prime might be expected round about $5 \cdot 10^{23}$.

It has been conjectured that the probability that a prime p is a Wilson prime is $1/p$.

Wolstenholme's numbers, and theorems

Joseph Wolstenholme (1829–1891) was professor of mathematics at the Royal Indian Engineering College. He was a friend of Leslie Stephen, whose daughter, Virginia Woolf, was a young girl when Wolstenholme shared their family holidays. She later incorporated Wolstenholme into one of her most famous books, *To the Lighthouse*, as the model for Mr. Augustus Carmichael.

In 1862 Wolstenholme proved that if p is a prime, not 2 or 3, then the numerator of the harmonic number,

$$1 + 1/2 + 1/3 + \ldots + 1/(p-1)$$

is divisible by p^2. The numerators of these sums are called the Wolstenholme numbers, and their sequence starts,

 1, 3, 11, 25, 137, 49, 363, 761, 7129, 7381, 83711, 86021, 1145993, 1171733, 1195757, 2436559, 42142223, . . . (Sloane A001008)

Similarly, the numerator of

$$1 + 1/2^2 + 1/3^2 + \ldots + 1/(p-1)^2$$

is divisible by p if and only if p is prime. (Alkan 1994) The sequence of these numerators starts,

 1, 5, 49, 205, 5269, 5369, 266681, 1077749, 9778141, 1968329, 239437889, . . . (Sloane A007406)

The numerator of $1 + 1/2^3 + 1/3^3 + \ldots + 1/(p-1)^3$ is divisible by p^2 if and only if p is prime and $p > 5$, and the numerator of $1 + 1/2^4 + 1/3^4 + \ldots + 1/(p-1)^4$ is divisible by p if and only if p is prime and $p > 7$.

Charles Babbage had noticed in 1819 that

$$\binom{2p-1}{p-1} \equiv 1 \pmod{p^2}$$

In 1862, Wolstenholme proved that

$$\binom{2p-1}{p-1} \equiv 1 \pmod{p^3}$$

For example, for $n = 7$:

$$13 \cdot 12 \cdot 11 \cdot 10 \cdot 9 \cdot 8 / 6 \cdot 5 \cdot 4 \cdot 3 \cdot 2 \cdot 1 = 1716 = 5 \cdot 7^3 + 1$$

The Wolstenholme converse—that if n satisfies the congruence then it is prime—has not been proved, but it is known to be true for even n and when n is a power of 3. (V. T. K. Weber, www.mat.unb.br)

If the same congruence is satisfied (mod p^4), then p is a Wolstenholme prime, but to date only two are known: 16843 and 2124679. (Sloane A088164) There are no other Wolstenholme primes less than $6 \cdot 4 \cdot 10^8$.

more factors of Wolstenholme numbers

Using Hisanori Mishima's World Integer Factorization Center site, which is described in the "Some Prime Web Sites" section of the bibliography, we find listed the following factors of the Wolstenholme numbers that are the numerators of the sums $1 + 1/2^2 + 1/3^2 + \ldots + 1/(n - 1)^2$. Only those lines are printed for which either $n + 1$ is a prime *or* $2n + 1$ is a prime, and in every case n or $2n + 1$, or both, divide the numerator:

n	numerator
2	5
3	7^2
4	$5 \cdot 41$
5	$11 \cdot 479$
6	$7 \cdot 13 \cdot 59$
8	$17 \cdot 63397$
9	$19 \cdot 514639$
10	$11 \cdot 178939$
11	$23 \cdot 43 \cdot 242101$
12	$13 \cdot 18500393$
14	$29 \cdot 7417 \cdot 190297$
15	$31 \cdot 37 \cdot 97 \cdot 1844659$
16	$17 \cdot 619 \cdot 78206663$
18	$19 \cdot 37 \cdot 8821 \cdot 38512247$
20	$41 \cdot 421950627598601$
21	$37 \cdot 43 \cdot 2621 \cdot 84786899$
22	$23 \cdot 295831 \cdot 52030193$
23	$47 \cdot 127 \cdot 31411862913089$

Continued on next page

26 $7 \cdot 53 \cdot 70853 \cdot 106357 \cdot 8408339$
28 $29 \cdot 7741 \cdot 46255855177282481$
29 $59 \cdot 6823 \cdot 889327 \cdot 24411224990171$
30 $31 \cdot 61 \cdot 4673 \cdot 621059 \cdot 1593520622137$
31 $43 \cdot 205883 \cdot 949932031764836381561$
33 $67 \cdot 76379 \cdot 2815507 \cdot 4335581 \cdot 539136331$
35 $71 \cdot 134417 \cdot 3532568757748095886123$
36 $37 \cdot 41 \cdot 73 \cdot 82163 \cdot 3707026238799632467$
39 $79 \cdot 107 \cdot 821 \cdot 27687173093 \cdot 240624262981001$
40 $41 \cdot 34033 \cdot 1260236851 \cdot 26302882946248703$
41 $83 \cdot 14173 \cdot 6611909133130699273072137371$
42 $43 \cdot 3049 \cdot 9479 \cdot 8944008062473011161194199$
44 $89 \cdot 137 \cdot 5023 \cdot 866961917 \cdot 387317256075298577081$
46 $47 \cdot 1123 \cdot 22152121 \cdot 341217893 \cdot 51579188214962371$
48 $97 \cdot 3389 \cdot 639066781 \cdot 64028661493 \cdot 3381178689557843$
50 $101 \cdot 269 \cdot 1451 \cdot 6583793 \cdot 120269976377958592837407551127$

Woodall primes

The Woodall numbers, named after H. J. Woodall, who published an account of them in 1917 written with **Cunningham**, are of the form, $n \cdot 2^n - 1$. Prime Woodall numbers are sometimes called **Cullen primes** of the second kind. The sequence of Woodall numbers starts, 1, 7, 23, 63, 159, 383, 895, . . .

A Woodall number is prime only when $n = 2, 3, 6, 30, 75, 81, 115, 123, 249, 362, 384, . . .$

It has been conjectured that almost all Woodall numbers are composite.

The largest known Woodall prime is $667071 \cdot 2^{667071} - 1$ (200,815 digits). It was discovered by Manfred Toplic and Yves Gallot in April 2002. (Caldwell, *Prime Pages*)

The prime $2^{521} - 1$ can also be written as $512 \cdot 2^{512} - 1$, making it both a Woodall prime and a **Mersenne prime**. (Dobb: Caldwell, *Prime Pages*)

zeta mysteries: the quantum connection

> Although the Riemann zeta-function is an analytic function with
> [a] deceptively simple definition, it keeps bouncing around
> almost randomly without settling down to some regular asymp-
> totic pattern. The Riemann zeta-function displays the essence of
> chaos in quantum mechanics . . . smooth, and yet seemingly
> unpredictable.
>
> —*M. C. Gutzwiller (1990, 377)*

The primes in general and the **Riemann hypothesis** in particular
might seem to represent the purest mathematics possible, so why
might there be any connection at all with physics? Yet that is just
what David Hilbert and George Pólya conjectured. As Pólya
explained to Andrew Odlyzko:

> I spent two years in Göttingen ending around the beginning of 1914. I tried
> to learn analytic number theory from Landau. He asked me one day: "You
> know some physics. Do you know a physical reason that the Riemann
> hypothesis should be true." This would be the case, I answered, if the non-
> trivial zeros of the zeta-function were so connected with the physical prob-
> lem that the Riemann hypothesis would be equivalent to the fact that all the
> eigenvalues of the physical problem are real.
>
> I never published this remark, but somehow it became known and it is
> still remembered. (Pólya 1982)

Pólya is referring to the fact that the eigenvalues of a symmetrical
matrix are real. If the non-trivial zeros of the zeta function are $\frac{1}{2} + ib_n$,
then the b_n would be the eigenvalues. Neither he nor Hilbert, who
made the same suggestion, had the slightest idea what the matrix
might be.

A connection appeared as the result of a chance meeting in 1973,
when Hugh Montgomery was reluctantly introduced to the great
Freeman Dyson, the English-born mathematician and physicist who is
famous for having reconciled the theories of Julian Schwinger and
Richard Feynman in quantum theory. Montgomery remarked that he
had been studying the zeros of the Riemann zeta function and he
mentioned a formula for their distribution, $1 - (\sin{(\pi r)}/(\pi r))^2$. Dyson
replied at once, to his surprise and Montgomery's, that this was the
density of the pair correlation of eigenvalues of a certain set of random matrices. (Sabbagh 2002, 134–36) As it happens, physicists have

a well-developed theory of the distribution of the eigenvalues of random matrices, which are used to model the energy levels of nuclei and other systems of particles.

Montgomery and Dyson never met again, never spoke again, yet this one chance encounter pointed to a potentially profound connection between quantum mechanics and number theory, which mathematicians and physicists have been exploring ever since. In particular, Andrew Odlyzko found through massive calculation that the zeros of the zeta function have the same pair correlation function as the eigenvalues of what are called GUE (Gaussian unitary ensemble) matrices. This result, inspired by Montgomery, is now known as the Montgomery-Odlyzko law.

It is entirely appropriate that George Pólya, who wrote several popular books on the role of analogy in mathematics, should have been the original inspiration for such a wonderful sequence of "connections"! This analogy has already benefited both physicists and number theorists. Physicists benefit because energy levels are usually hard to compute, whereas mathematicians have very efficient methods for calculating zeros of the zeta function.

There is another, related, connection. Quantum chaos studies the transitions from quantum mechanical systems, which are not chaotic, and classical Newtonian systems, which can be. It has led physicists to make predictions about the relationships between energy levels that can then be applied, by analogy, to the Riemann zeta function. Surprise! The predictions are confirmed!

And that's not all! In her paper "Quantum-like Chaos in Prime Number Distribution and in Turbulent Fluid Flows," A. M. Selvam claims, "Number theoretical concepts are intrinsically related to the quantitative description of dynamical systems of all scales ranging from the microscopic subatomic dynamics to macroscale turbulent fluid flows such as the atmospheric flows," and in particular that the prime numbers are analogous to eddies in turbulent fluid flows and that the frequencies of prime numbers follow quantum-like mechanical laws. She adds another analogy: "Roger Penrose discovered in 1974 the quasiperiodic Penrose tiling pattern. . . . The fundamental investigation of tilings which fill space completely is analogous to investigating the manner in which matter splits up into atoms and natural numbers split up into product of primes." (Selvam 2001)

In addition, the Riemann zeta function also shows up in the theory of Brownian motion and in the study of diffusion and percolation processes. Even this, as you might have guessed, isn't all. Xiao-Song Lin of the University of California has noticed an analogy between the Jones polynomial in knot theory and the Ihara-Hashimoto-Bass zeta function in graph theory. Inspired by the Montgomery-Odlyzko law, Xiao-Song Lin has "carried out some computer experiments on the Jones polynomial. Our data indicate that, quite likely, zeros of the Jones polynomial of alternating knots may obey certain statistical laws as well."

Little could an ancient Greek have realized as he read Euclid's presentation of the prime numbers in his *Elements*, in a painfully convoluted manner because the Greeks lacked a really efficient language and notation for the subject, that there might just possibly be a connection with the knots he used to tie his shoes and the prime numbers.

Of course, none of this current speculation actually proves how the zeros of the Riemann zeta function behave. It could be that we have just what physicists are accustomed to and just what mathematicians do not find ultimately satisfactory—a useful but imperfect model of reality.

And yet—Pythagoras claimed that the universe was made of numbers and Leopold Kronecker (1823–1891) claimed that "God made the integers and all the rest is the work of man." Who knows? Perhaps the world is even more cunningly constructed out of the *prime* numbers!

Appendix A: The First 500 Primes

2 3 5 7 11 13 17 19 23 29 31 37 41 43 47 53 59
61 67 71 73 79 83 89 97

101 103 107 109 113 127 131 137 139 149 151 157
163 167 173 179 181 191 193 197 199

211 223 227 229 233 239 241 251 257 263 269 271
277 281 283 293

307 311 313 317 331 337 347 349 353 359 367 373
379 383 389 397

401 409 419 421 431 433 439 443 449 457 461 463
467 479 487 491 499

503 509 521 523 541 547 557 563 569 571 577 587
593 599

601 607 613 617 619 631 641 643 647 653 659 661
673 677 683 691

701 709 719 727 733 739 743 751 757 761 769 773
787 797

809 811 821 823 827 829 839 853 857 859 863 877
881 883 887

907 911 919 929 937 941 947 953 967 971 977 983
991 997

1009 1013 1019 1021 1031 1033 1039 1049 1051 1061
1063 1069 1087 1091 1093 1097

1103 1109 1117 1123 1129 1151 1153 1163 1171 1181
1187 1193

1201 1213 1217 1223 1229 1231 1237 1249 1259 1277
1279 1283 1289 1291 1297

1301 1303 1307 1319 1321 1327 1361 1367 1373 1381
1399

1409 1423 1427 1429 1433 1439 1447 1451 1453 1459
1471 1481 1483 1487 1489 1493 1499

1511 1523 1531 1543 1549 1553 1559 1567 1571 1579
1583 1597

1601 1607 1609 1613 1619 1621 1627 1637 1657 1663
1667 1669 1693 1697 1699

1709 1721 1723 1733 1741 1747 1753 1759 1777 1783
1787 1789

1801 1811 1823 1831 1847 1861 1867 1871 1873 1877
1879 1889

1901 1907 1913 1931 1933 1949 1951 1973 1979 1987
1993 1997 1999

2003 2011 2017 2027 2029 2039 2053 2063 2069 2081
2083 2087 2089 2099

2111 2113 2129 2131 2137 2141 2143 2153 2161 2179

2203 2207 2213 2221 2237 2239 2243 2251 2267 2269
2273 2281 2287 2293 2297

2309 2311 2333 2339 2341 2347 2351 2357 2371 2377
2381 2383 2389 2393 2399

2411 2417 2423 2437 2441 2447 2459 2467 2473 2477

2503 2521 2531 2539 2543 2549 2551 2557 2579 2591
2593

2609 2617 2621 2633 2647 2657 2659 2663 2671 2677
2683 2687 2689 2693 2699

2707 2711 2713 2719 2729 2731 2741 2749 2753 2767
2777 2789 2791 2797

2801 2803 2819 2833 2837 2843 2851 2857 2861 2879
2887 2897

2903 2909 2917 2927 2939 2953 2957 2963 2969 2971
2999

3001 3011 3019 3023 3037 3041 3049 3061 3067 3079 3083 3089

3109 3119 3121 3137 3163 3167 3169 3181 3187 3191

3203 3209 3217 3221 3229 3251 3253 3257 3259 3271 3299

3301 3307 3313 3319 3323 3329 3331 3343 3347 3359 3361 3371 3373 3389 3391

3407 3413 3433 3449 3457 3461 3463 3467 3469 3491 3499

3511 3517 3527 3529 3533 3539 3541 3547 3557 3559 3571

Appendix B: Arithmetic Functions

The arithmetic functions, $d(n)$, $\sigma(n)$, and $\phi(n)$, from $n = 1$–80.

n	$d(n)$	$\sigma(n)$	$\phi(n)$
1	1	1	1
2	2	3	1
3	2	4	2
4	3	7	2
5	2	6	4
6	4	12	2
7	2	8	6
8	4	15	4
9	3	13	6
10	4	18	4
11	2	12	10
12	6	28	4
13	2	14	12
14	4	24	6
15	4	24	8
16	5	31	8
17	2	18	16
18	6	39	6
19	2	20	18
20	6	42	8
21	4	32	12
22	4	36	10
23	2	24	22
24	8	60	8
25	3	31	20
26	4	42	12
27	4	40	18
28	6	56	12
29	2	30	28
30	8	72	8
31	2	32	30
32	6	63	16
33	4	48	20
34	4	54	16

(Continued on next page)

n	$d(n)$	$\sigma(n)$	$\phi(n)$
35	4	48	24
36	9	91	12
37	2	38	36
38	4	60	18
39	4	56	24
40	8	90	16
41	2	42	40
42	8	96	12
43	2	44	42
44	6	84	20
45	6	78	24
46	4	72	22
47	2	48	46
48	10	124	16
49	3	57	42
50	6	93	20
51	4	72	32
52	6	98	24
53	2	54	52
54	8	120	18
55	4	72	40
56	8	120	24
57	4	80	36
58	4	90	28
59	2	60	58
60	12	168	16
61	2	62	60
62	4	96	30
63	6	104	36
64	7	127	32
65	4	84	48
66	8	144	20
67	2	68	66
68	6	126	32
69	4	96	44
70	8	144	24
71	2	72	70
72	12	195	24
73	2	74	72
74	4	114	36
75	6	124	40
76	6	140	36
77	4	96	60
78	8	168	24
79	2	80	78
80	10	186	32

Glossary

These explanations are in addition to those made at the end of the introduction. Many terms are explained simply by going to their entry; for example, **abundant number**, **perfect numbers**, or **quadratic residues**. Others will be found by going to the index, where the page number of an entry that explains the term is in **bold**. A few, such as the first three examples of notation in this glossary, are explained in the text, but it will not be obvious where.

Notation

$\phi(n)$ [some authors write phi(n)] is **Euler's totient function**, equal to the number of integers less than the integer n that are also prime to n, that is, have no common factor with n.

So $\phi(10) = 4$ because there are four numbers less than 10 and prime to it: 1, 3, 7, 9.

$d(n)$ means the number of divisors of the integer n. The unit, 1, is included and so is the number n, so $d(12) = 6$ because 12 has the divisors 1, 2, 3, 4, 6, and 12 itself.

$\sigma(n)$ is the sum of the divisors of the integer n, so $\sigma(12) = 1 + 2 + 3 + 4 + 6 + 12 = 28$.

Terms

aliquot parts The expression is rather old-fashioned. The aliquot parts of a number are its proper divisors, meaning its divisors apart from the number itself. The number 1 is always included. So the aliquot parts of 30 are 1, 2, 3, 5, 6, 10, and 15.

binomial coefficients These appear in algebra when a power such as $(1 + x)^5$ is expanded:

$$(1 + x)^5 = 1 + 5x + 10x^2 + 10x^3 + 5x^4 + x^5$$

The binomial coefficients in this case are 1, 5, 10, 10, 5, 1. They also appear in **Pascal's triangle**.

common factor Two integers have a common factor if a number, other than 1, divides both of them exactly.

composite number A number that can be written as the product of two or more prime numbers. The number 1 is neither a prime number nor a composite number.

coprime Two integers are coprime if they have no common factor.

letters Letters stand for integers unless otherwise indicated. The letter c often stands for a constant that may *not* be an integer.

logarithm The natural logarithm of n, the log to base e, is written as log n. This does *not* mean the usual logarithm to base 10, which would be written $\log_{10} n$.

number Throughout this book, "number" refers to a positive integer or whole number, unless stated otherwise.

primitive prime factor Given a sequence of integers, a primitive prime factor is a prime factor of a number in the sequence that is *not* a factor of any previous number in the sequence. In other words, that particular prime factor is appearing for the first time.

squarefree If any prime factor of an integer divides it more than once, then it is *not* squarefree. Otherwise it is squarefree. For example, $12 = 2^2 \cdot 3$, so 12 is not squarefree, and neither is $90 = 2 \cdot 3^2 \cdot 5$. However, $15 = 3 \cdot 5$ and $105 = 3 \cdot 5 \cdot 7$ are both squarefree.

Bibliography

The popular journals *American Mathematical Monthly,* and *Mathematics Magazine* will be found in many academic libraries, together with many professional journals in number theory.

Books marked with an asterisk are especially recommended (however, this does not imply that they are all elementary).

Abel, U., and H. Siebert (1993). Sequences with large numbers of prime values. *American Mathematical Monthly* 100(2):167–69.

Agrawal, M. (2003). Radio essay transcript and audio. www.boeing.com.

Agrawal, M., N. Kayal, and N. Saxena (2002). "PRIMES is in P." www.cse.iitk.ac.in/news.

Alkan, E. (1994). Variations on Wolstenholme's theorem. *American Mathematical Monthly* 101(10):1001–4.

Alspach, B. (2004). Cited in R. B. Eggleton and W. P. Galvin, Upper bounds on the sum of principal divisors of an integer. *Mathematics Magazine* 77(3):190–200.

Annapurna, U. (1938). Inequalities for $\sigma(n)$ and $\phi(n)$. *American Mathematical Monthly* 45(4):187–90.

Apostol, T. M. (2000). A centennial history of the prime number theorem. In R. P. Bambah, V. C. Dumir, and R. J. Hans-Gill, *Number theory.* Berlin: Birkhauser.

*Bach, E., and J. Shallit (1996). *Algorithmic number theory, vol. 1: Efficient algorithms.* Cambridge, MA: MIT Press.

Bailey, D. H., and J. M. Borwein (2000). Experimental mathematics: Recent developments and future outlook. Pp. 51–66 in *Mathematics unlimited— 2001 and beyond,* vol. 1, ed. B. Engquist and W. Schmid. New York: Springer.

Baker, A., B. Bollobás, and A. Hajnal (1990). *A tribute to Paul Erdös,* Cambridge, UK: Cambridge University Press.

Bass, T. A. (1995). Gene genie. *Wired Magazine* 3.08 (August 1995).

Baxa, C. (1993). A note on Diophantine representations. *American Mathematical Monthly* 100(2):138–43.

Beauregard, R. A., and E. R. Suryanarayan (2001). Square-plus-two primes. *Mathematical Gazette* 85(502):90–91.

*Beiler, A. (1966). *Recreations in the theory of numbers.* New York: Dover.

Bell, E. T. (1951). *Mathematics: Queen and servant of science*. London: Dover.

Benson, D. C. (1999). *The moment of proof*. Oxford, UK: Oxford University Press.

Berndt, B. C., and W. F. Galway (n.d.). On the Brocard-Ramanujan Diophantine equation n! + 1 = m, www.math.uiuc.edu.

Billingsley, P. (1973). Prime numbers and Brownian motion. *American Mathematical Monthly* 80(10):1099–1115.

Bing, W. Z., R. Fokkink, and W. Fokkink (1995). A Relation between partitions and the number of divisors. *American Mathematical Monthly* 102(4):345–47.

Bollobás, B. (1998). To prove and conjecture: Paul Erdös and his mathematics. *American Mathematical Monthly* 105(3):209–37.

Borwein, D., J. M. Borwein, P. B. Borwein, and R. Girgensohn (1996). Ginga's conjecture on primality. *American Mathematical Monthly* 103(1): 40–50.

Bencze, M. (1998). Problem 10655. *American Mathematical Monthly* 105(4):366.

Benito, M., and J. L. Varona (2001). www.loria.fr/~zimmerma/records/aliquot.html.

Brillhart, J., D. H. Lehmer, J. Selfridge, B. Tuckerman, and S. S. Wagstaff Jr. (1983/1988). Factorizations of $b^n \pm 1$, b = 2, 3, 5, 6, 7, 10, 11, 12 up to high powers. *Contemporary mathematics* 22. American Mathematical Society, Providence, RI, 3rd edition, 2002.

*Bressoud, D. M. (1989). *Factorization and primality testing*. New York: Springer.

Browkin, J. (2000). The *abc*-conjecture. In R. P. Bambah, V. C. Dumir, and R. J. Hans-Gill, *Number theory*. Berlin: Birkhauser.

*Burton, D. M. (1976). *Elementary number theory*. Boston: Allyn & Bacon.

Clarke, A. C., and G. Lee (1990). *Rama II*. New York: Bantam, p. 435.

Collison, M. J. (1980). The unique factorization theorem: From Euclid to Gauss. *Mathematics Magazine* 53(2):96–100.

*Conway, J. H., and R. K. Guy (1996). *The book of numbers*. New York: Springer.

Conway, J. H., R. K. Guy, W. A. Schneeberger, and N. J. A. Sloane (1997). The primary pretenders. *Acta Arithmetica* 78:307–13.

Cramér, H. (1936). On the order of magnitude of the difference between consecutive prime numbers. *Acta Arithmetica* 2:23–46.

Crandall, R., and C. Pomerance (2001). *Prime numbers: A computational perspective*. New York: Springer.

Creyaufmueller, W. (2002). www.loria.fr/~zimmerma/records/aliquot.html.

Crocker, R. (1961). A theorem concerning prime numbers. *Mathematics Magazine* 34(6):316, 344.

Crubellier, M., and J. Sip (1997). Looking for perfect numbers. Pp. 389–410 in *History of Mathematics: History of Problems*. Paris: Ellipses-Marketing.

Cunningham, A. J. C., and H. J. Woodall (1925). *Factorisations of $y^n - 1$, $y = 2, 3, 5, 6, 7, 10, 11, 12$ up to high powers.* London: Hodgson.

Deleglise, M., and J. Rivat (1996). Computing $\pi(x)$: the Meissel, Lehmer, Lagarias, Miller, Odlyzko method. *Mathematics of Computation* 65(123): 235–45.

*Derbyshire, J. (2003). *Prime obsession: Bernhard Riemann and the greatest unsolved problem in mathematics.* Washington, D.C.: Joseph Henry Press.

*Devlin, K. (2002). *The millennium problems.* New York: Basic Books.

*Dickson, L. E. (1952). *History of the theory of numbers,* 3 vols. New York: Chelsea Publishing Co.

Donnelly, H. (1973). On a problem concerning Euler's phi-function. *American Mathematical Monthly* 80(9):1029–31.

Dubner, H., and T. Forbes (2001). Prime Pythagorean triangles. *Journal of Integer Sequences* 4:1–11.

Dudeney, H. E. (1917). *Amusements in mathematics.* London: Thomas Nelson and Sons Ltd.

Dudley, U. (1978). *Elementary number theory,* 2nd ed. San Francisco: W. H. Freeman.

———. (1983). Formulas for primes. *Mathematics Magazine* 56(1):17–22.

———. (1994). Smith numbers. *Mathematics Magazine* 67:62–65.

Dusart, P. (1999). The kth prime is greater than k (ln k + ln ln k) for k ≥ 2. *Mathematics of Computation* 68(255):411–15.

Edwards, A. W. F. (1964). Infinite coprime sequences. *Mathematical Gazette* 48(366):416–22.

Edwards, H. M. (1974). *Riemann's zeta function.* New York: Academic Press.

Erdös, P. (1973). Über die Zahlen der Form sigma(n)-n und n-phi(n). *Elemente der Mathematik* 28:83–86.

Erdös, P., and M. Kac (1940). The Gaussian law of errors in the theory of additive number theoretic functions. *American Journal of Mathematics* 62:738–42.

Erdös, P., and U. Dudley (1983). Some remarks and problems in number theory related to the work of Euler. *Mathematics Magazine* 56(5):292–98.

Erdös, P., and C. V. Eynden (1992). Only finitely many rows in Pascal's triangle consist entirely of r-th-power-free integers, Problem E 3424. *American Mathematical Monthly* 99(6):579–80.

Euler, L. *Opera Omnia,* ser. 1, vol. 2, p. 241, quoted in Pólya 1954, p. 91.

Ferrier, A. (1947). *Les nombres premiers.* Paris: Librairie Vuibert.

Finch, S. R. (2003). *Mathematical Constants.* Cambridge, UK: Cambridge University Press.

Ford, K. (1999). The number of solutions of $\phi(x) = m$. *Annals of Mathematics,* series 2, 150(1):283–311.

Forman, R. (1992). Sequences with many primes. *American Mathematical Monthly* 99(6):555.

Franel, J., and E. Landau (1924). Les suites de Farey et le problème des nombres premiers. *Göttinger Nachrichten,* 198–206.

Frei, G. (1985). Leonhard Euler's convenient numbers, *Mathematical Intelligencer* 7(3):555–64.

Freudenthal, H. (1975). *Dictionary of scientific biography,* ed. C. C. Gillespie, entry, "Riemann, G. F. B.," New York: Charles Scribner's.

Friedlander, J., and H. Iwaniec (1998). The polynomial $x^2 + y^4$ captures its primes. *Annals of Mathematics* 148(3):945–1040.

Fung, G. W., and H. C. Williams (1990). Quadratic polynomials which have a high density of prime values. *Mathematics of Computation* 55(191): 345–53.

Gardiner, V., R. Lazarus, N. Metropolis, and S. Ulam (1956). On certain sequences of integers defined by sieves. *Mathematics Magazine* 29(3):117–22.

Gardner, M. (1975). *Martin Gardner's sixth book of mathematical diversions from "Scientific American."* New York: Scribner.

———. (1977). Mathematical games: A new kind of cipher that would take millions of years to break. *Scientific American* 237(August 1977):120–24.

*Gauss, C. F. (1801/1966). *Disquisitiones arithmeticae,* trans. Arthur A. Clarke. New Haven: Yale University Press.

———. (1817). *Göttingischegelehrie Anzeigen* for March 1817. Reproduced at http://newtuniv.home.att.net/index.html.

Gautschi, W., ed. (1995). *Mathematics of computation 1943–1993: A half-century of computational mathematics.* Providence, RI: American Mathematical Society.

Gethner, E., S. Wagon, and B. Wick (1998). A stroll through the Gaussian primes. *American Mathematical Monthly* 105(4):327–37.

*Glaisher, J. W. L., ed. (1940). *Number-divisor tables.* Cambridge, England: Cambridge University Press for the British Association.

Golomb, S. W. (1962). On the ratio of N to $\pi(N)$. *American Mathematical Monthly* 69(1):36–37.

Golomb, S. W. (1970). Powerful numbers. *American Mathematical Monthly* 77(8):848–52.

Golomb, S. (1981). The evidence for Fortune's conjecture. *Mathematics Magazine* 54(4):209–10.

Graham, R. L. (1964). A Fibonacci-like sequence of composite numbers. *Mathematics Magazine* 37(5):322–24.

Greaves, G. (2001). *Sieves in number theory.* New York: Springer.

Greger, K. (1978). Square divisors and square-free numbers. *Mathematics Magazine* 51(4):211–19.

Grimm, C. A. (1969). A conjecture on consecutive composite numbers. *American Mathematical Monthly* 76(10):1126–28.

Gupta, Shyam Sunder (n.d.). www.shyamsundergupta.com/amicable.htm.

Gutzwiller, M. C. (1990). *Chaos in classical and quantum mechanics.* New York: Springer.

*Guy, Richard K. (1981). *Unsolved problems in number theory.* New York: Springer.

————. (1983). Conway's prime producing machine. *Mathematics Magazine* 56(1):26–33.

————. (1988). The strong law of small numbers. *American Mathematical Monthly* 95(8):697–712.

————. (1990). The second strong law of small numbers. *Mathematics Magazine* 63(1):3–20.

*————. (1994). *Unsolved problems in number theory,* 2nd ed. New York: Springer.

————. (1997). Divisors and desires. *American Mathematical Monthly* 104(4):359–60.

Haimo, D. T. (1995). Experimentation and conjecture are not enough. *American Mathematical Monthly* 102(2):102–12.

Halberstam, H., and H-E. Richert (1974). *Sieve methods.* London: Academic Press, p. 3.

Harborth, H. (1977). Divisibility of binomial coefficients by their row number. *American Mathematical Monthly* 84(1):35–37.

Hardy, G. H. (1915). Prime numbers, in *Collected papers of G. H. Hardy,* vol. 2 (1967). Oxford, UK: Clarendon Press.

————. (1920). Some famous problems of the theory of numbers, in *Collected papers of G. H. Hardy,* vol. 2 (1967). Oxford, UK: Clarendon Press.

————. (1940). *Ramanujan: Twelve lectures on subjects suggested by his life and work.* Cambridge, England: Cambridge University Press.

————. (1941/1992). *A Mathematician's Apology.* Cambridge University Press.

————. (1966). Goldbach's theorem, in *Collected papers of G. H. Hardy,* vol. 1, pp. 545–60. Oxford, UK: Clarendon Press.

Hardy, G. H., and J. E. Littlewood (1923). Some problems of "partitio numerorum": III: On the expression of a number as a sum of primes, *Acta Mathematica* 1:1–70, reprinted in *Collected papers of G. H. Hardy,* vol. 1, pp. 561–630. Oxford, UK: Clarendon Press.

*Hardy, G. H., and E. M. Wright (1979). *An introduction to the theory of numbers,* 5th ed. Oxford, UK: Clarendon Press.

Hawkins, D. (1958). Mathematical sieves. *Scientific American* 199:105.

Hayes, B. (1998). *American Scientist* 86(2):113.

Hodges, L. (1993). A Lesser-known Goldbach conjecture. *Mathematics Magazine* 66(1):45–47.

*Hoffman, P. (1998). *The man who loved only numbers.* London: Fourth Estate.

Hoggatt, V. E., Jr., and M. Bicknell-Johnson (1977). Composites and primes among powers of Fibonacci numbers. *Fibonacci Quarterly* 15:2.

Honsberger, R. (1976). *Mathematical gems II.* Providence, RI: Mathematical Association of America.

Ingham, A. E. (1932/1990). *The distribution of prime numbers.* Cambridge, England: Cambridge University Press.

Kac, M. (1959). *Statistical independence in probability, analysis and number theory.* Carus Mathematical Monographs, no. 12. New York: Wiley.

———. (1987). *Enigmas of chance*. Berkeley: University of California Press.

Keller, W., and J. Richstein (2004). Solutions of the congruence $a^{p-1} \equiv 1$ (mod p^r). *Mathematics of Computation* (electronic only: posted June 8, 2004).

Kim, S. H., and C. Pomerance (1989). The probability that a random probable prime is composite. *Mathematics of Computation* 53:721–41.

Klee, V. (1969). Is There an n for which $\phi(x) = n$ has a unique solution? *American Mathematical Monthly* 76(3):288–89.

Knuth, D. E. (1981). *The art of computer programming*, vol. 2, *Seminumerical algorithms*. Reading, MA: Addison-Wesley.

———. (1990). A Fibonacci-like sequence of composite numbers. *Mathematics Magazine* 63(1):21–25.

Koshy, T. (2002). *Elementary number theory with applications*. San Diego: Harcourt/Academic Press.

Kumar, D., P. Ch. Ivanov, and H. E. Stanley (2003). Information entropy and correlations in prime numbers. Preprint. http://xxx.lanl.gov/abs/cond-mat/0303110.

Lagarias, J. C. (2002). An elementary problem equivalent to the Riemann hypothesis. *American Mathematical Monthly* 109:534–43.

Lal, M. (1967). Primes of the form $n^4 + 1$, *Mathematics of Computation* 21(98):245–47.

Laubenbacher, R. C., and D. J. Pengelley (1994). Gauss, Eisenstein, and the "third" proof of the quadratic reciprocity theorem: Ein kleines Schauspiel." *Mathematical Intelligencer* 16(2):67–68. In Wells 1997, 101–2.

Lauritzen, W. Versatile numbers-versatile economics. www.earth360.com/math-versatile.html.

Leech, J. (1957). Note on the distribution of prime numbers. *Journal of the London Mathematical Society* 32:56–58.

Lehmer, D. H. (1940). Introduction to *Number Divisor Tables*, ed. J. W. L. Glaisher. Cambridge, England: Cambridge University Press. [p. viii]

———. (1956). On the roots of the Riemann zeta-function. *Acta Arithmetica* 95:291–98.

Lemmermeyer, F. (2000). *Reciprocity laws: From Euler to Eisenstein*. London: Springer.

Mackay, A. L. (1994). *Dictionary of scientific quotations*. Philadelphia: Institute of Physics Publishing.

Mackenzie, D. (1997). Homage to an itinerant master. *Science* 275:759.

McLean, K. R. (2002). The prime factors of $2^n + 1$. *Mathematical Gazette* 86(507):466–67.

Makowski, A. (1960). On some equations involving functions $\phi(n)$ and $\sigma(n)$. *American Mathematical Monthly* 67(7):668–70.

Martin, G. (1999). The smallest solution of $\phi(30n + 1) < \phi(30n)$ is. . . . *American Mathematical Monthly* 106(6):449–51.

Menabrea, L. F. (1842). *Sketch of the Analytical Engine invented by Charles Babbage.* . . . www.fourmilab.ch/babbage/sketch.html.

Mollin, R. A. (1997). Prime-producing quadratics. *American Mathematical Monthly* 104(6):529–44.

*————. (2001). *An introduction to cryptography.* New York: Chapman and Hall/CRC.

Mollin, R. A., and P. G. Walsh (1986). On powerful numbers. *International Journal of Mathematics and Mathematical Sciences* 9(4):801–6.

Montgomery, P. L. (1993). New solutions of $a^{(p-1)} - 1 \equiv 1 \pmod{p^2}$. *Mathematics of Computation* 61(203):361–63.

Nash, C. (2000). Conjecture 21. www.primepuzzles.net/conjectures. conj_21.htm. (Carlos Rivera; Prime Puzzles & Problems Connection. Web site.)

Nelson, H. (1978–79). Problem 654. *Journal of Recreational Mathematics* 11.

Nelson, C., D. E. Penney, and C. Pomerance (1974). 714 and 715. *Journal of Recreational Mathematics* 7(2):87.

Newman, J. R. (1956). *The world of mathematics.* London: Simon & Schuster.

Newman, D. J. (1997). Euler's ϕ function on arithmetic progressions. *American Mathematical Monthly* 104(3).

Nicely, T. R. (2004a). Enumeration to $1.6 \cdot 10^{15}$ of the twin primes and Brun's constant. www.trnicely.net/counts.html.

————. (2004b). Enumeration to $1.6 \cdot 10^{15}$ of the prime quadruplets. www.trnicely.net/counts.html.

O'Connor, J. J., and E. F. Robertson (n.d.). *MacTutor History of Mathematics Archive.* www-history.mcs.st-andrews.ac.uk/HistTopics/Perfect_numbers. html.

Odlyzko, A. (1993). Iterated absolute values of differences of consecutive primes. *Mathematics of Computation* 61(203):373–80.

Odlyzko, A. M., and A. Schönhage (1988). Fast algorithms for multiple evaluations of the Riemann zeta function. *Transactions of the American Mathematical Society* 309:797–809.

*Ore, O. (1948/1988). *Number theory and its history.* Dover.

Papp, D., P. Jobling, G. Woltman, and Y. Gallot (2002). The chronology of prime number records. perso.wanadoo.fr/yves.gallot/primes.

Pasolov, V. V., and A. B. Sossinsky (1997). *Knots, links, braids and 3-manifolds,* Translations of Mathematical Monographs, vol. 154. Providence, RI: American Mathematical Society.

Poincaré, H. (n.d.). *Science and method.* New York: Dover.

Polignac, A. de (1849). Six propositions arithmologiques déduites de crible d'Érastosthène, *Nonvelle Annales de Mathématiques* 8:423–29.

*Pólya, G. (1954). *Mathematics and plausible reasoning,* vol. 1, *Induction and analogy in mathematics.* Princeton, NJ: Princeton University Press.

————. (1959). Heuristic reasoning in the theory of numbers. *American Mathematical Monthly* 66(5):375–84.

*————. (1962). *Mathematical discovery,* 2 vols. New York: Wiley.

————. (1982). Letter to Andrew Odlyzko, January 3, 1982. www.dtc. umn.edu/~odlyzko/polya/.

Pomerance, C. (1979). The prime number graph. *Mathematics of Computation* 33(145):399–408.

———. (1996). A tale of two sieves. *Notices of the AMS* 43(12).

Quine, W. V. (1988). Fermat's Last Theorem in combinatorial form. *American Mathematical Monthly* 95(7):636.

Ranganathan, S. R., ed. (1967). *Ramanujan, the man and the mathematician.* New Delhi: Asia Publishing House. (Recording the reminiscences of R. Srinivasan.)

Raimi, R. A. (1976). The first digit problem. *American Mathematical Monthly* 83(7):521–38.

Ramanujan, S. (1913). Irregular numbers. *Journal of the Indian Mathematical Society* 5:105–6.

———. (1915). Highly composite numbers. *Proceedings of the London Mathematical Society.* 2(XIV):347–409.

Ratering, S. (1991). An interesting subset of the highly composite numbers. *Mathematics Magazine* 64(5):343–46.

*Ribenboim, P. (1989). *The book of prime number records.* New York: Springer.

*———. (1991). *The little book of big primes.* New York: Springer. (A popular version of Ribenboim 1988.)

*———. (1995). *The new book of prime number records.* New York: Springer.

*Riesel, H. (1994). *Prime numbers and computer methods for factorization,* 2nd ed. Boston: Birkhäuser.

Rivera, C. (n.d.). Conjecture 4: Fermat primes are finite. The Prime Puzzles and Problems Connection. www.primepuzzles.net.

———. (n.d.). Conjecture 10: Champions and primorial numbers. The Prime Puzzles and Problems Connection. www.primepuzzles.net.

———. (n.d.). Conjecture 29: Sierpinski-like numbers; Carlos Rivera, Frank Buss. The Prime Puzzles and Problems Connection. www.primepuzzles. net.

———. (n.d.). Problem 29: Carlos Rivera, Eric Brier, Yves Gallot. The Prime Puzzles and Problems Connection. www.primepuzzles.net.

———. (n.d.). Problem 30: The Prime Puzzles and Problems Connection. www.primepuzzles.net.

———. (n.d.). Problem 44: Twin-primes producing polynomials race. The Prime Puzzles and Problems Connection. www.primepuzzles.net.

———. (n.d.). Puzzle 31: Carlos Rivera, Jo Yeong Uk, and Jack Brennen. The Prime Puzzles and Problems Connection. www.primepuzzles.net.

———. (n.d.). Puzzle 36: Carlos Rivera. The Prime Puzzles and Problems Connection. www.primepuzzles.net.

———. (n.d.). Puzzle 122: Consecutive twin primes. The Prime Puzzles and Problems Connection. www.primepuzzles.net.

Rouse Ball, W. W. (1939). *Mathematical Recreations and Essays.* New York: Macmillan.

Rubinstein, R. (n.d.). D. H. Lehmer's number sieves. http://ed-thelen. org/comp-hist/TheCompMusRep/TCMR-V04.html#Sieves.

Ruiz, S. M. (1997). A result on prime numbers. *Mathematical Gazette* 81(491):269–270.

*Sabbagh, K. (2002). Dr. Riemann's zeros. London: Atlantic.

Sacks, O. (1985). *The man who mistook his wife for a Hat, and other clinical tales*. New York: Simon & Schuster.

*Sautoy, M. de (2003). *The music of the primes*. London: Fourth Estate.

Schneider, W. (2002). Lucky numbers. www.wschnei.de.

———. (2003). Descriptive primes. www.wschnei.de.

Schroeder, M. R. (1992). The unreasonable effectiveness of number theory in physics, communication and music. In *The unreasonable effectiveness of number theory*, ed. S. A. Burr. Providence, RI: American Mathematical Society.

Selvam, A. M. (2001). Quantum-like chaos in prime number distribution and in turbulent fluid flows. In the Canadian electronic journal *APEIRON* 8(3):29–64.

———. (n.d.). Quantum-like chaos in prime number distribution and in turbulent fluid flows. Indian Institute of Tropical Meteorology, Pune 411 008, India. www.geocities.com/amselvam/prime3/Prime3.html.

Shor, P. W. (1997). Polynomial-time algorithms for prime factorization and discrete logarithms on a quantum computer. *SIAM J. Computing* 26:1484–1509.

Seife, C. (2002). Erdös's hard-to-win prizes still draw bounty hunters. *Science* 296:39–40.

Silverman, J. H. (1988). Wieferich's criterion and the abc-conjecture. *Journal of Number Theory* 30(2):226–37.

Singmaster, D. (1983). Some Lucas pseudo-primes. *Abstracts of the AMS*, quoted in Ribenboim (1995), p. 104.

*Sloane, N. J. A. (1973). *A handbook of integer sequences*. New York: Academic Press.

*Sloane, N. J. A., and S. Plouffe (1995). *The encyclopedia of integer sequences*. Academic Press, San Diego, California.

Stein, M. L., S. M. Ulam, and M. B. Wells (1964). A visual display of some properties of the distribution of primes. *American Mathematical Monthly* 71:516–20.

Stephens, N. M. (1971). On the Feit-Thompson conjecture. *Mathematics of Computation* 25:625.

Subrakamian, P. K., and S. F. Becker (1966). On highly composite consecutive integers. *American Mathematical Monthly* 73(6):510–13.

Sugden, A. M. (2001). Biological model generates prime numbers. *Science* 293(5528).

Vaughan, R. C. Chance News 11.02, Chance in the primes: Chapter 2. www.dartmouth.edu/chance/chance_news/recent_news/chance_primes _chapter2.html.

Weinstein, L. (1966). Divisibility properties of Fibonacci numbers. *Fibonacci Quarterly* 4(1):83.

Weisstein, E. W. (2001). Consecutive number sequences. *MathWorld.* mathworld.wolfram.com/ConsecutiveNumberSequences.html.

———. (n.d.). Goldbach conjecture. *MathWorld.* mathworld.wolfram.com/GoldbachConjecture.html.

*Wells, D. G. (1986). *The Penguin dictionary of curious and interesting numbers.* London: Penguin.

———. (1992). *The Penguin book of curious and interesting puzzles.* London: Penguin.

———. (1997). *The Penguin book of curious and interesting mathematics.* London: Penguin.

Wilf, H. S. (1987). The editor's corner. *American Mathematical Monthly* 94(1):3–6.

Willans, C. P. (1964). On formulae for the Nth prime number. *Mathematical Gazette* 48.

Williams, H. C., and J. O. Shallit (1994). Factoring integers before computers. In W. Gautschi, ed., *Mathematics of computation 1943–1993: A half-century of computational mathematics.* Providence, RI: American Mathematical Society.

Williams, H. C. (1996). *Mathematics of Computation* 61(203).

*———. (1998). *Éduoard Lucas and primality testing.* New York: Wiley.

Wolf, M. (1996). On the twin and cousin primes. http://rose.lft.uni.wroc.pl/~mwolf.

Wright, E. M. (1951). A Prime-representing function. *American Mathematical Monthly* 58(9):616–18.

Young, R. M. (1998). Probability, pi, and the primes. *Mathematical Gazette* 82(495):443–51.

Zwillinger, D. (1979). A Goldbach conjecture using twin primes. *Mathematics of Computation* 33(147):1071.

Some Prime Web Sites

There are thousands of sites featuring prime numbers. Two enthusiasts vie, in my judgment, for the top Prize for Primes: Chris Caldwell and Neil Sloane.

Chris Caldwell is Professor in the Department of Mathematics and Statistics at the University of Tennessee at Martin. He got his PhD in algebraic number theory in 1984 and now specializes in number theory and using computers to teach mathematics—especially on the World Wide Web via his massive *Prime Pages* site, which has received more than a dozen awards for excellence.

The Prime Pages is at www.utm.edu/research/primes, where you will also find his personal Web site. A prominent icon of the Statue of Liberty at the top of his own site is labeled "America, a Beacon, not a Policeman," and there are links to www.againstbombing.org/index.htm, which is in turn titled, "Americans Against Bombing: Americans Against World Empire." I mention this both to show my approval and because, I am happy to note, mathematicians tend to be pacific and internationalist—perhaps for the obvious reason that mathematics itself crosses every kind of boundary, internally and externally.

The Prime Pages is an incredibly varied cornucopia of facts, figures, and references, proofs, puzzles, and conjectures, definitions and theorems, lists and tables—if you want it, there's a good chance Chris Caldwell has it.

But even he cannot have everything: the prime numbers are far too large a subject. So another dazzling site that is a pleasure to visit is Neil **Sloane's** *On-Line Encyclopedia of Integer Sequences*, at www.research.att.com/~njas/sequences.

"Integer Sequences" in the title might seem a bit limiting, but it's amazing how much mathematics is—or can be—linked to sequences of whole numbers. For example, **Wolstenholme's** theorems are about the sums of the harmonic series, that is, fractions, not integers, but of course fractions have numerators and denominators, and these form integer sequences! So type in "Wolstenholme," choose the option "word," and you get four references to the index,

Wolstenholme numbers: A001008, A007406, A007408, A007410

and twenty-one references to the database, starting with this lengthy entry:

A093689 Wolstenholme's theorem states that prime $p > 3$ divides A007406($p - 1$). It is not difficult to show that this implies p also divides A007406(($p - 1$)/2). In most instances, $a(n) = (\text{prime}(n) - 1)/2$. Exceptions occur for primes in A093690, which have a smaller $a(n)$. Note that if p divides A007406(k) for $k < (p - 1)/2$, then p divides A007406($p - k - 1$). Another interesting observation: it appears that $p = 7$ is the only prime that divides A007406(k) for some $k > p - 1$; 7 divides A007406(26) = 23507608254234781649. Also note that when $p > 3$ and $2p - 1$ are both prime, they divide A007406 ($p - 1$).

You can enter the consecutive terms of a sequence (you don't have to start at the beginning, which indeed is sometimes ambiguous) or a search term, such as "Wolstenholme," as we have seen, or a Sloane sequence number: the current catalog numbers all start with an A. (They used to start with M or N. The prime number sequence is now A00040; it used to be M0652 or N0241.)

MathWorld, at http://mathworld.wolfram.com, describes itself as "the web's most extensive mathematics resource." It is "Created and maintained by Eric Weisstein with contributions from the world mathematics community," and is supported by Wolfram Research and the National Science Foundation.

When I last checked, there were 11,838 entries, of which more than 260 are listed under "Prime Numbers."

An especially attractive feature is the *MathWorld* Headline News, which is currently reporting (July 15, 2004) that R. F. Arenstorf seems close to settling the **twin primes** conjecture and that Josh Findley has discovered the 41st **Mersenne prime.**

Walter Schneider at Kreuzmattenstrasse 8, 79276 Reute, Germany, runs *Mathews: The Archive of Recreational Mathematics* Web site, at www.wschnei.de/index.html.

Carlos Rivera is the author of *The Prime Puzzles & Problems Connection,* at www.primepuzzles.net. Readers will note several references in my bibliography. A nice feature is that it contains extensive responses from solvers.

The Wikipedia online encyclopedia also has interesting information about prime numbers, at http://en.wikipedia.org/wiki/Prime_number.

Hidden among Chris Caldwell's *Prime Pages* is a page by Andrew Booker, *The Nth Prime Page,* which allows you to find the Nth prime, up to $N = 10^{12}$, or the value of $\pi(N)$, for N up to $3 \cdot 10^{13}$.

Hisanori Mishima runs the WIFC (World Integer Factorization Center) at www.asahi-net.or.jp/~KC2H-MSM/mathland/matha1, which includes listings of factors of many types of numbers from **Cunningham** to **Fermat** to **Riesel** to **Woodall** and many in between.

Yves Gallot, who is a prolific finder of assorted primes, factorizations etc., has his own *Chronology of Prime Number Records* at http://perso.wanadoo.fr/yves.gallot/primes/chrrcds.html.

Dario Alpern has a useful site at www.alpertron.com.ar that has one good feature and three brilliant features. You can solve any quadratic equation in two variables; you can use the "Factorization using the Elliptic Curve Method" facility, which gives not just the factors of the number you type in but also $d(n)$, the number of **divisors**, $\sigma(n)$, their sum, and $\phi(n)$, **Euler's totient function;** you can use his "Discrete logarithm calculator," which allows you to solve $a^x \equiv b \pmod{N}$ for x; and you can play with his **Ulam spiral** applet, which gives the equation of the diagonal line through your mouse-selected point, its coordinates, and the first few values of the function. Excellent!

Index